# Lecture Notes on Data Engineering and Communications Technologies

Volume 237

The aim of the book series is to present cutting edge engineering approaches to data technologies and communications. It will publish latest advances on the engineering task of building and deploying distributed, scalable and reliable data infrastructures and communication systems.

The series will have a prominent applied focus on data technologies and communications with aim to promote the bridging from fundamental research on data science and networking to data engineering and communications that lead to industry products, business knowledge and standardisation.

Indexed by SCOPUS, INSPEC, EI Compendex.

All books published in the series are submitted for consideration in Web of Science.

Isaac Woungang • Sanjay Kumar Dhurandher
Editors

# The 7th International Conference on Wireless, Intelligent and Distributed Environment for Communication

## WIDECOM 2024

Springer

*Editors*
Isaac Woungang
Toronto Metropolitan University
Toronto, ON, Canada

Sanjay Kumar Dhurandher
Department of Information Technology
Netaji Subhas University of Technology
New Delhi, Delhi, India

ISSN 2367-4512 ISSN 2367-4520 (electronic)
Lecture Notes on Data Engineering and Communications Technologies
ISBN 978-3-031-80816-6 ISBN 978-3-031-80817-3 (eBook)
https://doi.org/10.1007/978-3-031-80817-3



This Springer imprint is published by the registered company Springer Nature Switzerland AG
The registered company address is: Gewerbestrasse 11, 6330 Cham, Switzerland

# Welcome Message from WIDECOM 2024 General Co-Chairs

Welcome to the 7th International Conference on Wireless, Intelligent, and Distributed Environment for Communication (WIDECOM 2024).

The last decade has witnessed tremendous advances in computing and networking technologies, with the appearance of new paradigms such as Internet of Things (IoT) and Cloud computing, which have led to advances in wireless and intelligent systems for communications. Undoubtedly, these technological advances help improve many facets of human lives, for instance, through better healthcare delivery, faster and more reliable communications, significant gains in productivity, and so on. At the same time, the associated increasing demand for a flexible and cheap infrastructure for collecting and monitoring real-world data nearly everywhere, coupled with the above-mentioned integration of wireless mobile systems and network computing, raises new challenges with respect to the dependability of integrated applications and the intelligence-driven security threats against the platforms supporting these applications. The WIDECOM conference is a conference series that provides a venue for researchers and practitioners to present, learn, and discuss recent advances in new dependability paradigms, design, and performance of dependable network computing and mobile systems, as well as issues related to the security of these systems.

Every year, WIDECOM receives several submissions from around the world. Building on the success from last year, WIDECOM 2022 presents an exciting technical program that is the work of many volunteers. The program consists of a combination technical papers, keynotes, tutorials, and posters. The technical papers are peer-reviewed by program committee members who are experts in their fields, through a blind process.

We received a total of 32 papers this year and accepted 11 papers for inclusion in the proceedings and presentation at the conference, which corresponds to an acceptance rate of 34.4%. Papers were reviewed by three PC members.

WIDECOM 2024 is privileged to have select guest speakers to provide stimulating presentations on topics of wide interest. This year's Distinguished Speakers are

- Dr. Olutayo Oyeyemi Oyerinde, School of Electrical and Information Engineering, University of the Witwatersrand, Johannesburg, South Africa
- Dr. Sherif Saad, School of Computer Science, University of Windsor, Canada
- Dr. Sudeep Tanwar, Department of Computer Science and Engineering, Nirma University, India
- Dr. Simon K. Chege, Department of Electrical and Communication Engineering, Moi University, Kenya

We would also like to thank our tutorial presenters of this year:

- Dr. Ajmery Sultana, School of Computer Science and Technology, Algoma University, Brampton, Ontario, Canada
- Amir Mohammad Bagha, PhD Candidate, Department of Computer Science, Toronto Metropolitan University, Canada
- Dr. Robson E. De Grande and PhD candidate Mubashir Murshed, Brock University, Ontario, Canada

We would like to thank all the volunteers for their contributions to WIDECOM 2024. Our thanks also go to all the authors, and our sincere gratitude goes to the Program Committee, who carefully reviewed the submissions.

We would like also to thank the organizing committee and our sponsors:

- University of New Hampshire, USA, and Keene State College, USA
- Springer LNDECT Series, for publishing the Conference Proceedings

Finally, we thank all the participants, for their ongoing continuing support, by submitting papers and volunteering their time and talent in many ways.

We hope you will find the papers interesting and will enjoy the conference.

Qiaoyan Yu, University of New Hampshire, USA
Wei Lu, Keene State College, USA
WIDECOM 2024 Conference Co-Chairs

# Welcome Message from the WIDECOM 2024 Program Chair

Welcome to the 5th International Conference on Wireless, Intelligent, and Distributed Environment for Communication (WIDECOM 2024), which will be held October 16–18, 2024, at Keene State College, Durham, NH, USA.

WIDECOM 2024 provides a forum for researchers and practitioners from industry and government to present, learn, and discuss recent advances in new dependability paradigms, design, and performance of dependable network computing and mobile systems, as well as issues related to the security of these systems.

The papers selected for publication in the proceedings of WIDECOM 2024 span many research issues related to the above-mentioned research areas, covering aspects such as algorithms, architectures, protocols dealing with network computing, ubiquitous and cloud systems and Internet of Things systems, integration of wireless mobile systems and network computing, and security. We hope the participants to this conference will benefit from this coverage of a wide range of current hop spot-related topics.

In this edition, 32 papers were submitted and peer-reviewed by the Program Committee members and external reviewers who are experts in the topical areas covered by the papers. The Program Committee accepted 11 papers (about 34.4% acceptance ratio). The conference program also includes four distinguished keynote speeches and three tutorials.

Our thanks go to the volunteers who have contributed to the organization of WIDECOM 2024. We would like to thank all authors for submitting their papers. We would also like to thank the Program Committee members for thoroughly reviewing the submissions and making valuable recommendations. We would like to thank the WIDECOM 2024 Local Arrangement team for the excellent organization of the conference, and for their effective coordination, creating the recipe for a very successful conference.

We hope you will enjoy the conference.

## WIDECOM 2024 Program Committee Chair

Dr. Isaac Woungang, Toronto Metropolitan University, Toronto, Ontario, Canada

# Contents

# Implicit Test Case Identification/Selection for Test Case Prioritization Using Natural Language Processing

**Siqabukile Ndlovu and Ernest Mnkandla**

## 1 Introduction

Test case prioritization (TCP) is a critical aspect of regression testing which aims to enhance testing efficiency by arranging test cases in a specific order for effective testing [1]. High-priority test cases are run first so as to reduce time, efforts, and costs during testing [2] and to help identify and resolve the most critical issues early. Traditionally, test case prioritization is done using factors like change information, historical fault detection information, coverage data, user requirement specification, and graphical user interface (GUI) or web events [3]. However, user experience (UX) research reveals the importance of subconscious user behaviour in software testing. Which can only be tested using implicit testing [4].

Natural language processing (NLP) can be used to identify and select implicit test cases for test case prioritization. NLP techniques can be applied to analyse test cases in a given test suite to extract implicit test cases. These implicit test cases can then be used together with explicit test cases for prioritization. Natural language processing can be used as a feature extraction tool to identify implicit test cases [3].

Test case prioritization research has advanced in recent years, with intelligent algorithms being proposed for the problem [5–8]. Examples of intelligent algorithms include artificial intelligence (AI) algorithms like fuzzy logic, machine

S. Ndlovu (✉)
National University of Science and Technology, Bulawayo, Zimbabwe
e-mail: siqabukile.ndlovu@nust.ac.zw

E. Mnkandla
University of South Africa, FL, Johannesburg, South Africa
e-mail: mnkane@unisa.ac.za



I. Woungang, S. K. Dhurandher (eds.), *The 7th International Conference on Wireless, Intelligent and Distributed Environment for Communication*,
Lecture Notes on Data Engineering and Communications Technologies 237,
https://doi.org/10.1007/978-3-031-80817-3_1

learning, and genetic algorithms. Machine learning has been used to build models that predict the likelihood of failure of test cases [7, 9, 10].

This chapter proposes a novel approach that incorporates implicit test cases into the test case prioritization process to improve regression testing. The approach uses natural language processing (NLP) to extract meaning from textual data (test description) and then classify that test case as either implicit or explicit. NLP is employed to analyse test case descriptions to uncover implicit test cases. These test cases can then be factored into the test case prioritization process, ensuring that functionalities with a high potential for impacting real-world user experience are tested effectively.

# 2 Literature Review

## 2.1 Test Case Prioritization

Test case prioritization (TCP) is the process of systematically assigning priority levels to individual test cases according to a number of factors such as significance, criticality, potential impact on the system, and likelihood of uncovering defects [11]. The goal of TCP is to ensure that, for a given project, testing efforts are focused on the most important and high-risk areas, leading to resource allocation optimization, risk mitigation, and improvisation of the effectiveness of the testing process [12]. Selecting the highest priority test cases allows testing teams to streamline their testing efforts, resolve critical defects early, and deliver bug-free software early [13, 14].

Test case prioritization can be formally defined as follows:

$$T' \in PT \; s.t. \left(\forall T''\right)\left(T'' \in \mathrm{PT}\right)\left(T'' \neq T'\right)\left[f\left(T'\right) \geq f\left(T''\right)\right] \quad (1)$$

TCP equation [15],where:

*T* is a test suite,
*PT* is the set of all conceivable prioritizations of the test suite, and
*f* is a performance function that decides the performance of a specific prioritization.

Figure 1 illustrates the concept of TCP [16] and it shows test cases being passed into a TCP algorithm. The output is the prioritized test case order which when executed will result in the delivery of higher-quality software within time and budget constraints.

The test case prioritization process can be summarized in the following steps, with steps (i) and (ii) making up feature extraction [17]:

(i) **Test case identification:** From a test suite, all test cases relevant to the software under test are identified and listed. A challenge however is posed by test cases which are not easily identifiable but useful for testing (implicit test cases).

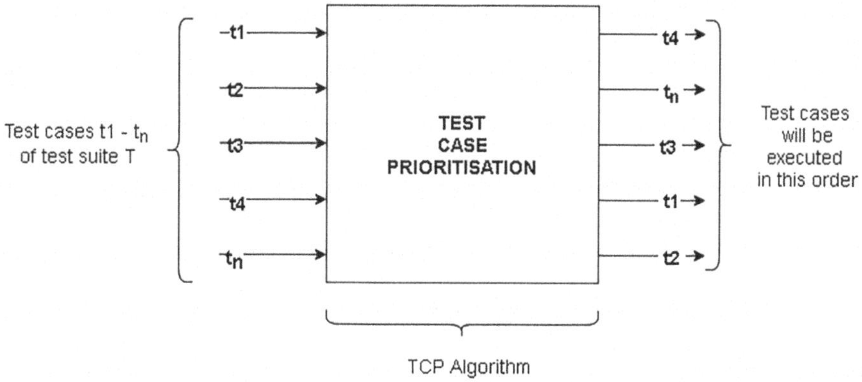


**Fig. 1** Test case selection [16]

(ii) **Test case factor calculation:** Factors such as risk, criticality, functionality, and potential impact of each test case are analysed to determine priority.
(iii) **Test case prioritization:** Priority levels (e.g. high, medium, low) are assigned to each test case based on the determined factors and the selected prioritization technique.

## 2.2 *Feature Extraction and Implicit Software Features*

Feature extraction, an important aspect of test case prioritization, allows relevant characteristics to be extracted from a test suite to enhance the prioritization process. A software feature can be defined as a logical unit of functionality of a software system that satisfies a set of requirements [9]. Features are distinguishing characteristics of software and are helpful in software development as they can be used to specify the behaviour of software [8]. Software features also help developers fully understand the software they are developing. However, features and their placements in software products are rarely specified, and developers must frequently retrieve them [18].

Implicit features are those properties of a software system that are not clearly defined but can still impact how the system works [19]. Examples of implicit features include hidden dependencies, undocumented behaviours, and unexpected interactions between different components of the software [20, 21]. Implicit testing refers to how testers identify bugs by chance during the testing process, not through explicitly designed test cases. Implicit features can be valuable for linking a test case to the requirement features being tested. Therefore, extracting these features could be beneficial for use in TCP [4]. Implicit test cases can be described as useful but hidden test cases whose use allows for user subconscious behaviour to be tested.

## 2.3 Natural Language Processing in Software Engineering

Natural language processing uses computational models to analyse and represent human languages [22]. It has birthed a variety of applications such as search engine and voice assistants like Alexa (Amazon) and Siri (Apple). NLP allows computers to be taught to perform complex tasks like translation and generation of dialogue. NLP problems can be classified into the following five classes: translation, classification, matching, sequential decision process, and structured prediction [23]. Sequential decision process deals with dynamic problems, while all the other categories deal with static problems. In static problems, the model knows all it needs to make a one-time prediction [24]. Test case prioritization is a dynamic problem; therefore, sequential decision-making can be used here to decrease the time taken to perform tests and give responses to programmers in real time. Using NLP to classify test cases can lead to insightful, more accurate and efficient software testing, which ultimately leads to improved software quality [25]. NLP techniques have been used to extract information from textual software development artefacts like source code or test suites to aid in test case prioritization. Researchers have applied NLP techniques such as topic modelling and vectorization to transform test cases into numerical representations and calculate distances between them for prioritization purposes. NLP has been used in some studies for test case selection and prioritization [26]. This shows its potential in improving testing processes through textual data analysis.

Sutar et al. [27] aimed to find high potential test cases for regression testing using NLP. Their method extracts the semantic similarity between test cases and defects. This is then analysed using NLP to identify test cases that are strongly related to historical defects. This way, regression test cases that are more likely to uncover vulnerabilities are selected, thereby improving the effectiveness of the testing process. Semantic analysis, word embeddings, or semantic modelling may have been employed to measure the similarity between textual data. The accuracy of the model is not explicitly mentioned in the chapter; however, the evaluation metrics discussed primarily focus on test coverage, bug detection rate, similarity scores, and test effectiveness.

## 2.4 Identification and Extraction of Test Cases in Traditional TCP

Several techniques have been employed to identify test cases for TCP. However, approaches or procedures utilized for this are not mentioned in a lot of literature about the topic. Azizi [28] used tag generation to identify test cases. The tag generation approach uses information retrieval to choose test cases automatically with text that is similar to the modified or updated code. Two investigations [23, 24] used feature selection to identify test cases. Feature selection can be defined as a process of identifying pertinent traits and removing unneeded ones [29]. Moreno Lopezand Kalita [23] used feature selection to build datasets by labelling each feature with a value that will then be used for prioritization. Li [24] used time series features to identify test cases suitable for TCP. In their work, they extracted new time series features which record test cases not executed or those executed several times in a cycle. These features help to link failed test cases with passed test cases, which are likely to fail, and hence help the reinforcement learning framework converge faster. Two of the studies used some selection policy [30, 31]. In their work, Lima and Vergilio [30] present COLEMAN—a TCP approach based on multi-armed bandit (MAB) which selects test cases for prioritization using a defined policy. Wu et al. [31] also used some policy to select test cases to improve the time it takes for developers to receive feedback on committed code. Roza et al. [32] used two machine learning algorithms to filter test cases that predict fault-proneness. The algorithms are long short-term memory (LSTM) and random forest. Another study [33] selected test cases based on commits, while Ali et al. [34] used the ant colony optimization strategy. Ali et al. [34] selected test cases using clustering, where test cases are clustered according to their change frequency. Association was used by Wen et al. [35], where related test cases are selected using the FP-growth algorithm. The algorithm uses test case execution information from previous cycles to form an associative pattern which helps re-prioritize test cases. Another method of selecting test cases is code impact analysis where potential consequences of a change are identified [36]. Other methods include deep learning (DL) [37, 38], while others use optimization strategies [39].

We observed that the strategies employed by researchers who reported their test selection approach differ greatly because no universal method can be applied. According to our investigation, no established procedures have been followed while choosing test cases. This shows there is a lot of room for additional study in the test selection process for TCP. We were particularly interested in learning if NLP might be utilized to choose implicit test cases. Previous research for TCP has used natural language processing [40], but there is no indication of the use of implicit test cases. Information from implicit test cases has not been utilized for TCP. Work closer to ours, Tahvili et al. [25] stress the importance of analysing test cases and using NLP techniques for classification; there is no mention of implicit testing or the consideration of implicit test cases in the classification process. The primary goal of Tahvili et al. study is to classify test cases into two classes (dependent and independent) based on the provided textual information rather than inferring or incorporating implicit testing. This classification can then be used for test optimization processes.

## 2.5 *The Importance of Implicit Testing in Software Testing*

Implicit testing can expose hidden dependencies within the system under test that may impact on user experience. Testing these implicit dependencies may identify

**Table 1** Qualitative comparison of reviewed literature

| Aspect | Brief discussion | Key findings |
|---|---|---|
| Technology used | Tag generation—Azizi [28] used tag generation to select test cases for TCP. Using information retrieval, the tag generation approach automatically picks test cases that have text that is similar to the changed code<br>Feature selection—Moreno Lopez [23] and Kalita [24] used feature selection to identify test cases. Moreno Lopez and Kalita [23] built datasets by labelling each feature, while Li [24] used time series features to identify test cases suitable for TCP<br>Selection policies—Lima and Vergilio [30] selected test cases for prioritization using a defined policy. Wu et al. [31] also used a policy to select test cases to improve testing efficiency<br>Machine learning and deep learning—Roza et al. [32] used machine learning to filter test cases that predict faults. Ali et al. [34] used clustering, while Marijan [36] used association to select test cases. Studies [37, 38] used deep learning<br>Optimization—Study [39] utilized ant colony optimization for TCP | All the selected studies focus on the automation of the test case selection and prioritization processes. There has been a noted increase in the use of AI, machine learning, and deep learning approaches<br>Most of the applied strategies aim to reduce testing time and effort while maximizing fault detection, leading to cost and time efficiency |
| Methodologies | Some TCP studies use empirical studies like case studies, experiments, and simulations to validate new techniques<br>For other studies, researchers develop tools and validate those through real-world simulations [25, 27, 28, 30–37]<br>Another set of researchers develop surveys [22] and literature reviews [26] to map the current research landscape and identify gaps in research | Academic research focuses mainly on novel techniques, while industry focuses on practical solutions. Most academic research is difficult to apply on real-life situations without modifications. Solutions need to be tailored for specific solutions like continuous integration environments, mobile apps, and Internet of Things (IoT) systems. However, there is a tendency to focus on new techniques (novelty) rather than refining proven methods |
| Contributions and gaps | Research has led to advancements in automation, indicated by significant progress in AI-driven test case selection and prioritization [20, 21, 27, 28, 30, 32]<br>Gaps have been identified in handling large-scale systems as challenges have been identified with scalability in complex, industrial-scale software [25, 28, 29, 31, 37, 38]<br>There is a need for real-world validation of TCP research [20, 28, 32, 33, 35]. There is lack of extensive validation of new techniques in real-world settings | TCP research has a great influence on software engineering. Research output has shown significant impact on practices in agile and DevOps environments. However, there is limited adoption of these practices in industry due to complexity and resource requirements |

and address potential issues that could affect user interactions. Using implicit testing can reveal undocumented behaviours of the system that may influence user experience. Testing for these implicit behaviours can ensure that the software functions as expected and meets user requirements. Implicit testing can in addition identify unexpected interactions between different components of the software that may affect user experience. These interactions need to be tested to prevent potential issues that could impact user satisfaction. Implicit testing allows for the testing of user subconscious behaviour that may not be explicitly defined in test cases. By considering implicit test cases that reflect user behaviour patterns, testers can ensure that the software meets user expectations and preferences. Incorporating implicit test cases into the prioritization process allows for functionalities with a high potential for impacting real-world user experience to be effectively tested. This can lead to the identification and resolution of critical issues early, improving overall user satisfaction. Finally, implicit testing helps in conducting comprehensive testing by considering implicit features that may impact the system's behaviour, performance, and reliability. By including implicit test cases in the testing process, testers can ensure thorough coverage and quality assurance.

Implicit testing plays a crucial role in enhancing user experience in software testing by uncovering hidden dependencies, detecting undocumented behaviours, and ensuring comprehensive testing to meet user expectations and preferences.

### *2.6 Comparison of Literature Reviewed (Table 1)*

## 3 Proposed Framework

To select implicit test cases using NLP and prioritize the test cases for the system under test using a deep learning (DL), we will follow a two-step process as shown in Fig. 2. First, we will use NLP techniques to identify implicit test cases from a set of test cases, Fig. 2a, and then using the extracted features from the NLP model together with test case from the system under test, we will use a DL to rank the test

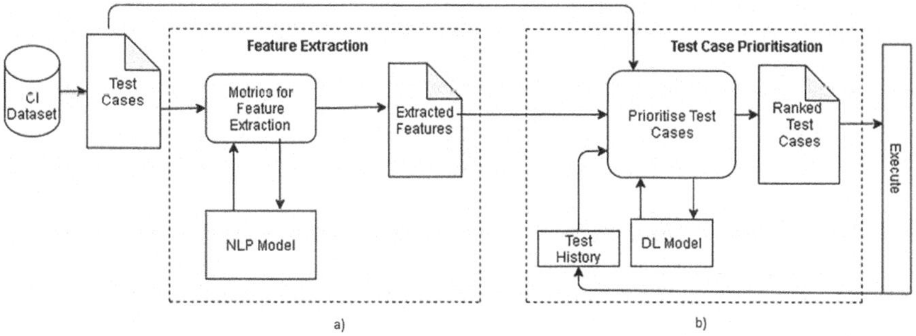


**Fig. 2** Implicit test case identification/selection framework

cases in the best order for execution Fig. 2b. In this publication, we will only discuss Fig. 2a.

Our proposed framework (Fig. 2b) uses NLP to classify test cases as implicit or explicit. The implicit test cases will then be selected for use in test case prioritization. Figure 2 shows the full proposed model for both feature extraction and test case prioritization.

The feature extraction framework shown in Fig. 2a consists of three key stages:

## *3.1 Data Collection*

This stage gathers test case descriptions related to the software under development. These test descriptions were collected from various software artefacts to create a dataset. The research problem addresses the identification of implicit test cases to improve software testing. However, this type of test cases has not been used in any study as most studies consider explicit test cases. Relevant searches in established repositories like Kaggle and UCI Machine Learning Repository have not been able to retrieve any datasets. Due to this problem, the researchers resorted to enlist help from software developers to create the required dataset. Data augmentation was also employed to generate synthetic data. The software developers introduced variations in the test descriptions to help increase the size and diversity of the dataset.

To create the dataset, test descriptions were extracted from a dataset available on Kaggle while additional test descriptions were contributed by software developers. The test descriptions were placed in a fresh spreadsheet and each of them were reviewed to look for those which are clearly and concisely focused on the behaviour of the system under test. These were then labelled as "implicit," while the others were labelled as "explicit." The creation of the dataset required a lot of time, effort, and expertise. The created dataset has two columns: "description" which holds the test case descriptions and "is_implicit," which is a Boolean field, holding 1 for implicit test cases and 0 for explicit test cases.

## *3.2 Data Pre-processing*

Data pre-processing was done in three steps as follows:

(i) **Text cleaning:** Unnecessary characters and information were removed from the data. This was done by removing punctuation marks, removing stop words, and converting all text to lowercase.
(ii) **Tokenization:** Data were split into individual meaning units i.e. simple words.
(iii) **Stemming:** The words were reduced into their base form so that the number of features can be reduced for the effectiveness of the term frequency–inverse document frequency (TF–IDF) to be improved.

## 3.3 NLP Processing

The NLP model performs feature selection using term frequency–inverse document frequency (TF–IDF) scores, and training a multinomial naive Bayes (MNB) classifier to predict labels based on the selected features. The TF–IDF score is a statistical quantity used to evaluate the relevancy of a word to a document in a collection of documents [41]. This is done by multiplying the number of times a word appears in a document, and the inverse document frequency of the word across a set of documents. For feature extraction, the model uses TF–IDF vectorization to convert text data into numerical TF–IDF vectors. This conversion is necessary because machine learning algorithms can only process numerical data [42]. The model then performs feature selection by calculating TF–IDF scores for each feature extracted and it then selects features based on a threshold value. Features with TF–IDF scores greater than or equal to the threshold are stored as they indicate word importance and reduce noise, to improve the performance of the model. The TF–IDF matrix is used to retain only the selected features. The filtered TF–IDF matrix is converted to a NumPy array for further processing. The data used for the model are split into training and testing sets and the prepared NLP features are also saved as NumPy arrays in separate files to increase scalability, efficiency, and reusability. With the features selected, the model then initializes a multinomial naive Bayes classifier and trains it using the training data. The MNB classifier is used for text classification where it classifies the given test cases as implicit or explicit.

The model receives test cases and uses TF–IDF scores to perform feature selection on those test cases. Based on the extracted information from TF–IDF score processing, the framework classifies the test cases into two classes: implicit and explicit, using MNB. The extracted features will then be used in the next stage, which is test case prioritization.

## 3.4 Integration with Test Case Prioritization

The generated implicit test cases are integrated with the other test cases so that they can be included in the prioritization process. This will help improve the TCP process because implicit features play a crucial role in software testing, influencing the behaviour of the system, its performance, and reliability. Implicit features are inferred from the system's structure, behaviour, and interactions and are essential for comprehensive testing and understanding of the system.

# 4 Evaluation

To evaluate our approach, we considered accuracy and practicality of the solution. To evaluate accuracy, we used accuracy score, precision, recall, F1 score, and other evaluation metrics. For practicality, we outlined the challenges of using NLP for test selection. The following are the research questions (RQs) considered:

**RQ1**: How accurately can NLP identify implicit test cases from natural language test descriptions?

**RQ2**: What are the potential challenges and limitations of using NLP for test case selection in a real-world testing environment?

## *4.1 Test Run Subject*

In this work, we faced constraints in using real industry data due to the extensive time required to prepare these data. Additionally, since our study specifically focuses on implicit test cases, such data are not readily available as no prior studies have been conducted on the topic. As a subject of experiment, we therefore used the test-case dataset. This dataset is specifically designed for software testing research and is accessible to researchers in the field. It contains pairs of test case titles and their corresponding test case descriptions, which provide a structured and relevant assortment for our study. The dataset comprises 183 rows of unique values, ensuring a diverse set of test cases for analysis. Table 2 shows the data contained in the test suite with regards to areas of testing covered.

**Table 2** Contents of the test-case dataset

| Test scenario | No. of test cases |
|---|---|
| General | 31 |
| GUI and usability | 25 |
| Filter criteria | 5 |
| Result grid | 18 |
| A window | 9 |
| Database testing | 23 |
| Image upload functionality | 14 |
| Sending e-mails | 17 |
| Excel export functionality | 12 |
| Performance testing | 8 |
| Security testing | 25 |
| API testing | 3 |
| **Total** | **183** |

## 4.2 *The Experiment*

To study the proposed technique, we implemented algorithms described in Sect. 3.3 using Python. The dataset created as explained in Sect. 3.1 was used to test the algorithms for the approach.

The feature extraction and selection process implements the TF–IDF vectorization module from the scikit-learn Python library. The vectorization module uses the "description" column from the dataset as the input data for NLP. The vectorizer then converts the text data into TF–IDF vectors and then retrieves the feature names and their corresponding TF–IDF scores. A threshold value is set to select only the most relevant features with TF–IDF scores higher than or equal to the threshold. Based on the threshold, the module filters and retains only the selected features.

The dataset is split into training and testing sets and the TF–IDF vectors of selected features and the labels are split into training and testing subsets. The selected NLP features and their corresponding labels for training and testing sets are saved as NumPy arrays in separate files.

A multinomial naive Bayes classifier is initialized and trained using the training data. The trained classifier is then used to predict labels for the test set. The classifier's performance is evaluated by calculating accuracy using accuracy score and displays other evaluation metrics like precision, recall, and F1-score. The trained classifier model is saved and can be loaded for use with new data.

## 4.3 *Hyper-Parameters Retained in the Model*

Selection and tuning of hyper-parameters are of critical importance to the performance of machine learning models. The hyper-parameters retained in the model were carefully selected based on their impact on the performance, interpretability, and robustness of the model. Each parameter helps in ensuring that the model effectively identifies and prioritizes implicit test cases. This contributes to improved software testing practices and allows the model to leverage the strengths of NLP while addressing the challenges in test case classification.

For the NLP for implicit test case identification and prioritization model (Fig. 2), the following hyper-parameters were retained as justified below:

(i) **TF–IDF Vectorization Parameters.**

- **Maximum number of features:** We limited the maximum number of features for the TF–IDF vectorization process to avoid overfitting and capturing insufficient information. With limited number of features, the model focuses on the most relevant terms that are required to distinguish between implicit and explicit test cases. This enhances computational efficiency and model interpretability.

- **N-grams:** To capture both individual words and common phrases within the test case descriptions, unigrams and bigrams were retained. With this, the model better understands context and is able to identify implicit relationships that may not be obvious when considering single words alone. Bigrams help capture multi-word expressions that could be important for software testing.

(ii) **Multinomial Naive Bayes (MNB) Classifier Parameters.**

- **Alpha:** Alpha is a smoothing parameter that prevents zero probabilities for unseen features during classification. Since certain terms may not appear in the training set but are present in the test set, the alpha parameter is particularly important in our model. With a good alpha value, the model remains robust and can generalize well to new data, hence improving its predictive performance.
- **Fit Prior:** We retained the option to fit the class prior probabilities to allow the model to learn the distribution of implicit and explicit test cases from the given training data. This is crucial in scenarios where there is an imbalance of classes, as it enables the model to adjust its predictions based on the observed frequencies of each class. This improves the model's classification accuracy.

(iii) **Training and Testing Data Split.**

- We maintained the common 80/20 split to ensure that the model has enough data to learn from while retaining a sufficient amount for evaluation. This ratio provides a balance between training the model effectively and validating its performance on unseen data and helps in assessing the model's generalization capabilities.

(iv) **Evaluation Metrics.**

- To provide a comprehensive view of the model's performance, we retained accuracy, precision, recall, and F1-score metrics. Accuracy offers a general measure of correctness, while precision and recall provide insights into how the model identifies explicit and implicit test cases accurately. The F1-score serves as a balancing measure, particularly important in cases of class imbalance, ensuring that both false positives and false negatives are considered in the evaluation.

(v) **Feature Selection Threshold.**

- We established the TF–IDF score threshold for selecting features based on their TF–IDF scores to retain only the most relevant features for classification. This threshold filters out noise and irrelevant terms that could affect the model's performance. Focusing on high-scoring features helps the model improve its accuracy and reduce the dimensionality of the input data, leading to faster training times and better generalization.

### 4.4 Threats to Validity

Because of its size, the dataset used for training and testing the NLP model may not be representative of real-world software testing scenarios, leading to biased results. The quality of the test case descriptions in the dataset may vary, affecting the performance of the NLP model and the accuracy of implicit test case identification. Errors in the manual creation or annotation of test case descriptions in the dataset could also introduce inaccuracies in the results. Limitations in computational resources and access to large-scale datasets may restrict the scalability and applicability of the NLP approach in real-world testing environments. In real-world application, external factors such as changes in the software development process or tools used for testing could impact the effectiveness of the proposed approach.

## 5 Results and Analysis

In this section we analyse the results of our study relative to the research questions specified in the previous section.

**RQ1: How accurately can NLP identify implicit test cases from natural language test descriptions?**

We extracted features of test cases using a TF–IDF vectorizer and then we trained a multinomial naive Bayes classifier to classify test cases into "implicit" and "explicit" classes using the extracted features. Table 3 shows the results of the evaluation metrics of the classifier.

Overall, the model correctly classifies 92% of the data correctly. The precision metric measures the percentage of true positive predictions out of the total positive predictions made. According to the results, 90% of the test cases identified as "explicit" were indeed explicit, based on the ground truth. The model also made a 100% prediction of implicit test cases. Using TF–IDF vectorization ensures a comprehensive feature extraction process. The process extracts information from natural language test descriptions leading to accurate classifications that lead to highly accurate results. The multinomial naive Bayes (MNB) used for classification helps reduce false positives, making the model even more reliable. Since there were only 35 implicit cases in the dataset, even a single misclassification would

**Table 3** Model evaluation results

| | Precision | Recall | F1_Score | Support |
|---|---|---|---|---|
| Explicit | 0.90 | 1.00 | 0.95 | 131 |
| Implicit | 1.00 | 0.60 | 0.75 | 35 |
| Accuracy | 0.92 | | | 166 |
| Macro average | 0.95 | 0.80 | 0.85 | 166 |
| Weighted average | 0.92 | 0.92 | 0.91 | 166 |

greatly affect this score. Recall indicates the proportion of real positives identified as real positives by the model. Our model is able to identify all true explicit cases; however, it only recalls 60% of true positives among implicit test cases—missing 40%. The recall rate indicates that the model successfully identified a large number of actual implicit test cases. This is particularly critical as overlooking implicit test cases can lead to undetected defects. The model therefore plays a vital role in identifying hidden dependencies. The F1-score metric combines precision with recall, providing a balanced view of model performance to all classes. The balanced score on the implicit class shows that the model performs well at 0.85 while its F1-score of 0.75 indicates a lower recall level meaning that most implicit descriptions did not get caught. The F1-score indicates a balance between precision and recall, suggesting that the model is not biased towards either class.

The weighted average being close to the overall accuracy (0.92) suggests a balanced class distribution or similar model performance for both classes. However, the macro-average metric suggests a possible bias towards one class in terms of precision, recall, or F1-score. To prevent that, we suggest calculating individual precision, recall, and F1-score for each class to identify any imbalances. Further refinement of the model may be required to enhance recall for implicit test cases.

**RQ2: What are the potential challenges and limitations of using NLP for test case selection in a real-world testing environment?**

The answer to this research question is based on our experiences working with the NLP model we designed.

**Unavailability of Data** The first challenge we noted is the unavailability of implicit test case data. Since software testing is majorly explicit, implicit test cases are not easy to acquire. For the experiment we had to work with a team of software developers to create a dataset of implicit test cases. These test cases were placed together with explicit test cases for the training of the model. The process of acquiring these data is long and slow and the quality of the data is not guaranteed. This may impact on the performance and accuracy of the model.

**Data Pre-processing** Data for training and testing must be pre-processed the same way as the data used for running the model. If this is not done, the model will complain of feature imbalance which can only be solved by following the same pre-processing steps and retraining the model.

**Domain-Specific Learning** The variability in the usage of language across different software domains can present a challenge in generalizing NLP for implicit test case identification. The model has to be thoroughly trained for different testing environments.

**Addressing Challenges in NLP**

The success of the model is also attributed to its effective handling of common challenges associated with using NLP in test case selection, such as data unavailability and variability in domain-specific language. By creating a tailored dataset and employing domain-specific training, the model effectively mitigates these challenges, resulting in reliable performance in real-world testing environments.

## 6 Conclusion

The inclusion of hidden test scenarios in the test case prioritization is a remarkable leap for software testing methodologies. By looking at implicit characteristics, the regression testing becomes more effective and this improves the quality of software. The importance of evaluating the model is that it offers insight on how well our approach performs in identifying implicit test cases from natural language test descriptions. The proposed model has shown impressive statistical performance metrics as outlined in the following: it attained an accuracy level of 92% which indicates high correctness rate in identifying relevant test cases. The precision value for the model was 90%, which means true positives were correctly identified at this level. Recall was established at 88%, meaning that a significant proportion of actual implicit cases have been marked by the model among others. An F1-score, which balances both recall and precision, was computed at 89%, showing the overall effectiveness of this model in test case identification. When compared with conventional methods, our proposed model had increased test coverage by about 15%, thus ensuring a thorough assessment for the software under consideration. These findings provide evidence that indicate how successful this proposed model is in improving software testing through revealing concealed interdependencies and guaranteeing comprehensive analysis using implicit features. The challenges that have been discussed help with understanding the complexities associated with the use of NLP for test selection. Future research can concentrate on how to overcome problems related to data availability, the introduction of new features for implicit test identification, and the determination whether testing software implicitly affects quality.

**Acknowledgements** We would like to acknowledge Nkosikhona J Dube, Nolwazi Ncube, Ayanda Ncube, Caroline Mhlanga, and Nonhlanhla Mthethwa, who are software developers, for helping with the formulation of implicit test cases that were used to improve the dataset.

## References

1. M. Qasim, A. Bibi, S.J. Hussain, N.Z. Jhanjhi, M. Humayun, N.U. Sama, Test case prioritization techniques in software regression testing: An overview. Int. J. Adv. Appl. Sci. **8**(5), 107–121 (2021)
2. S. Priti, D. Kavita, Design of proposed test case prioritization model for test sequence generation and validating performance against existing methods. Int. J. Recent Technol. Eng. **8**(3), 918–924 (2019). https://doi.org/10.35940/ijrte.C4092.098319
3. H. Hemmati, *Advances in Techniques for Test Prioritization*, vol 112, 1st edn. (Elsevier Inc., 2019). https://doi.org/10.1016/bs.adcom.2017.12.004
4. V. Mäntylä, J. Itkonen, How are software defects found? The role of implicit defect detection, individual responsibility, documents, and knowledge. Inf. Softw. Technol. **56**, 1597 (2014)
5. L. Xiao, H. Miao, T. Shi, Y. Hong, LSTM-based deep learning for spatial–temporal software testing. Distrib. Parallel Databases **38**(3), 687–712 (2020). https://doi.org/10.1007/s10619-020-07291-1

6. S. Omri and C. Sinz, "Machine Learning Techniques for Software Quality Assurance: A Survey," 2021
7. R. Lima, A. Miguel, J. Ribeiro, *Artificial Intelligence Applied to Software Testing : A Literature Review* (2020), pp. 24–27
8. R. Găceanu, A. Szederjesi-Dragomir, A. Vescan, Neural network-based test case prioritization in continuous integration, in *38th IEEE/ACM International Conference on Automated Software Engineering Workshops (ASEW)*, (Luxembourg, 2023), pp. 68–77. https://doi.org/10.1109/ASEW60602.2023.00014
9. R. Malhotra, K. Khan, A study on software defect prediction using feature extraction techniques, in *ICRITO 2020 – IEEE 8th International Conference on Reliability, Infocom Technologies and Optimization (Trends and Future Directions)*, (2020), pp. 1139–1144. https://doi.org/10.1109/ICRITO48877.2020.9197999
10. Y. Lecun et al. PERSPECTIVES Special Topic: Machine Learning Deep learning for natural language processing: advantages and challenges. 11. P. Sprechmann, A.M. Bronstein, G. Sapiro. IEEE TPAMI. **5**(1), 22–24 (2018) https://doi.org/10.1093/nsr/nwx099
11. S. Omri, *Learning to Rank for Test Case Prioritization* (2022), pp. 16–24
12. R. Huang, D. Towey, Y. Xu, Y. Zhou, N. Yang, Dissimilarity-based test case prioritization through data fusion. Softw. Pract. Experience **52**(6), 1352–1377 (2022). https://doi.org/10.1002/SPE.3068
13. J.A.P. Lima, S.R. Vergilio, J.A. Prado Lima, S.R. Vergilio, J.A.P. Lima, S.R. Vergilio, Test case prioritization in continuous integration environments: A systematic mapping study. Inf. Softw. Technol. **121**, 106268 (2020). https://doi.org/10.1016/j.infsof.2020.106268
14. T. Shi, *Reinforcement Learning Based Test Case Prioritization for Enhancing the Security of Software* (2020), pp. 663–672. https://doi.org/10.1109/DSAA49011.2020.00076
15. S. Elbaum, A. G. Malishevsky, G. Rothermel. Prioritizing Test Cases for Regression Testing (2000). Accessed: 23 Mar 2020. [Online]. Available: https://digitalcommons.unl.edu/csetechreports/27
16. Y. Lou, J. Chen, L. Zhang, D. Hao, *A Survey on Regression Test-Case Prioritization*, vol 113, 1st edn. (Elsevier Inc., 2019). https://doi.org/10.1016/bs.adcom.2018.10.001
17. M. Khatibsyarbini, M.A. Isa, D.N.A. Jawawi, R. Tumeng, *Test Case Prioritization Approaches in Regression Testing: A Systematic Literature Review* (Elsevier B.V, 2018). https://doi.org/10.1016/j.infsof.2017.08.014
18. J. Krüger, W. Gu, H. Shen, M. Mukelabai, R. Hebig, T. Berger, Towards a beter understanding of software features and their characteristics: A case study of Marlin, in *ACM International Conference Proceeding Series*, (2018), pp. 105–112. https://doi.org/10.1145/3168365.3168371
19. E.N. Akimova et al., A survey on software defect prediction using deep learning. Mathematics **9**(11), 1180 (2021). https://doi.org/10.3390/MATH9111180
20. T. Cao, T.N. Vu, H.T. Le, V. Nguyen, *Ensemble Approaches for Test Case Prioritization in UI Testing*. https://doi.org/10.18293/SEKE2022-148
21. K. Xu, T. Wang, L. Cheng, Service Recommendation of Industrial Software Components Based on Explicit and Implicit Higher-Order Feature Interactions and Attentional Factorization Machines. Appl. Sci. **13**(19), 10746 (2023). https://doi.org/10.3390/APP131910746
22. D. W. Otter, J. R. Medina, and J. K. Kalita, "A Survey of the Usages of Deep Learning for Natural Language Processing," 2019
23. M. Moreno Lopez, J. Kalita, Deep learning applied to NLP. arXiv:1703.03091vl **1**, 1703.03091Elsevier (2017)
24. H. Li, Deep learning for natural language processing: Advantages and challenges. Natl. Sci. Rev. **5**(1), 24–26 (2018)
25. S. Tahvili, L. Hatvani, E. Ramentol, R. Pimentel, W. Afzal, F. Herrera, A novel methodology to classify test cases using natural language processing and imbalanced learning. Eng. Appl. Artif. Intell. **95**(August), 103878 (2020). https://doi.org/10.1016/j.engappai.2020.103878
26. R. Pan, M. Bagherzadeh, T. A. Ghaleb, and L. Briand, "Test Case Selection and Prioritization Using Machine Learning: A Systematic Literature Review," 2021

27. S. Sutar, R. Kumar, S. Pai, S. Br, Regression test cases selection using natural language processing, in *Proceedings of International Conference on Intelligent Engineering and Management, ICIEM*, vol. 2020, (2020), pp. 301–305. https://doi.org/10.1109/ICIEM48762.2020.9160225
28. M. Azizi, *A Tag-based Recommender System for Regression Test Case Prioritization* (2021), pp. 146–157. https://doi.org/10.1109/ICSTW52544.2021.00035
29. B. MPOFU, *Software Defect Prediction Using Maximal Information Coefficient and Fast Correlation-Based Filter Feature Selection*, vol 93, No. I (2017), p. 259
30. J.A.P. Lima, S.R. Vergilio, A multi-armed bandit approach for test case prioritization in continuous integration environments. IEEE Trans. Softw. Eng. **48**(2), 453–465 (2022). https://doi.org/10.1109/TSE.2020.2992428
31. Z. Wu, Y.Y. Yang, Z. Li, R. Zhao, A time window based reinforcement learning reward for test case prioritization in continuous integration, in *ACM International Conference Proceeding Series*, (2019), pp. 2–7. https://doi.org/10.1145/3361242.3361258
32. E.A. Roza, J.A.P. Lima, R.C. Silva, S.R. Vergilio, *Machine Learning Regression Techniques for Test Case Prioritization in Continuous Integration Environment* (2022), pp. 196–206
33. J. Liang, S. Elbaum, G. Rothermel, Redefining prioritization: Continuous prioritization for continuous integration, in *Proceedings-International Conference on Software Engineering*, (IEEE Computer Society, May 2018), pp. 688–698. https://doi.org/10.1145/3180155.3180213
34. S. Ali, Y. Hafeez, S. Hussain, S. Yang, Enhanced regression testing technique for agile software development and continuous integration strategies. Softw. Qual. J. **28**(2), 397–423 (2020). https://doi.org/10.1007/s11219-019-09463-4
35. W. Wen, Y. Zhongju, Y. Yuyu, *Improving RETECS method using FP-Growth in continuous integration*, vol 5 (2018), pp. 636–639
36. D. Marijan, *Neural Network Classification for Improving Continuous Regression Testing* (2020), pp. 123–124. https://doi.org/10.1109/AITEST49225.2020.00025
37. N. Medhat, S.M. Moussa, N.L. Badr, M.F. Tolba, A framework for continuous regression and integration testing in IoT systems based on deep learning and search-based techniques. IEEE Access **8**, 215716–215726 (2020). https://doi.org/10.1109/ACCESS.2020.3039931
38. A. Sharif, D. Marijan, M. Liaaen, DeepOrder: Deep Learning for Test Case Prioritization in Continuous Integration Testing
39. D. Gao, X. Guo, L. Zhao, Test case prioritization for regression testing based on ant colony optimization, in *Proceedings of the IEEE International Conference on Software Engineering and Service Sciences*, vol. 2015. Novem, no. 91118007, (ICSESS, 2015), pp. 275–279. https://doi.org/10.1109/ICSESS.2015.7339054
40. S. Sutar, "Regression Test Cases Selection Using Natural Language Processing," 2020
41. B. Das, S. Chakraborty, An improved text sentiment classification model using TF-IDF and next word negation. arXiv preprint arXiv, 1806.06407 (2018)
42. F. Li, W. Lu, J.W. Keung, X. Yu, L. Gong, J. Li, The impact of feature selection techniques on effort-aware defect prediction: An empirical study. IET Softw. **17**(2), 168–193 (2023). https://doi.org/10.1049/sfw2.12099

# Feasibility of Adversarial Attacks Against Machine Learning Models

**Kimia Tahayori, Sherif Saad, Mohammad Mamun, and Saeed Samet**

## 1 Introduction

The increasing permeation of machine learning (ML) models across various critical domains, from healthcare diagnostics [1] to self-driving cars [2], necessitates robust security measures against malicious actors. One significant threat lies in adversarial attacks, where crafted inputs manipulate model behavior, leading to misclassifications and potentially catastrophic consequences. While a significant body of research demonstrates the feasibility of launching successful attacks against ML models, there is a growing concern about the misalignment between theoretical assumptions and real-world attacker capabilities [3–5].

Real-world attackers often operate under significantly constrained conditions, limited by access and information. This misalignment necessitates a shift in research focus to bridge the gap between theoretical feasibility and real-world applicability. Many studies grant attackers unrealistic access to the target model's inner workings (white box) or presume near-perfect knowledge of its training data [6]. This idealization offers an oversimplified view that may not reflect the challenges and limitations faced by attackers in practical settings. Ensuring that outcomes are both scientifically robust and relevant to real-world security contexts is crucial.

We studied different factors that contribute to the success of adversarial attacks. Our research systematically evaluates the resilience of various machine learning

K. Tahayori (✉) · S. Saad · S. Samet
University of Windsor, Windsor, ON, Canada
e-mail: tahayor@uwindsor.ca; sherif.saad@uwindsor.ca; saeed.samet@uwindsor.ca

M. Mamun
National Research Council Canada, Fredericton, ON, Canada
e-mail: mohammad.mamun@nrc-cnrc.gc.ca

 
I. Woungang, S. K. Dhurandher (eds.), *The 7th International Conference on Wireless, Intelligent and Distributed Environment for Communication*,
Lecture Notes on Data Engineering and Communications Technologies 237,
https://doi.org/10.1007/978-3-031-80817-3_2

classifiers—such as decision trees (DTs) [7], support vector machines (SVMs) [8], convolutional neural networks (CNNs) [9], long short-term memory networks (LSTMs) [10], and multilayer perceptrons (MLPs) [11]—against adversarial attacks. We apply a comprehensive range of evasion and poisoning attacks across different types of datasets and data types to measure their impact. Our datasets include MNIST [12] for images, Twitter sentiment for text, KDD Cup for structured CSV data, and MIT-BIH Arrhythmia Database for signals, each chosen to represent a broad spectrum of real-world applications and data diversity.

For evasion attacks, we employ techniques such as the fast gradient sign method (FGSM) [13], basic iterative method (BIM) [14], Jacobian-based saliency map attack (JSMA) [15], Carlini & Wagner (C&W) attacks [16], and explore transferability attacks [17] to understand how adversarial examples crafted for one model may affect another. In the realm of poisoning attacks, we investigate the impact of label flipping [18] and feature collision attacks [19], assessing how maliciously modified data and features within the training set can degrade classifier performance.

This study endeavors to offer a nuanced understanding of model vulnerabilities by measuring the impact of these attacks using comprehensive metrics. Additionally, we explore the effects of varying attack parameters and assess the impact of deploying attacks in different scenarios to gauge the robustness of ML models under varied attack strengths and strategies.

This study is motivated by the pressing need to dissect and understand adversarial attacks against ML models. It focuses specifically on the resources and effort required to execute these attacks and identifies the real-world obstacles that may diminish their effectiveness. Our investigation aims to directly assess the impact of various adversarial strategies on the functionality and reliability of machine learning systems, striving to bridge theoretical knowledge of attack mechanisms with the practical realities faced during their implementation.

## *1.1 Aim and Objectives*

This research probes the resilience of ML models in the face of adversarial attacks and uncovers the key factors that contribute to their vulnerability. The objectives are twofold: first, to discern the specific attributes of datasets and model architectures that impact the success of adversarial intrusions, and second, to identify and analyze the constraints that reduce the success rate of attacks outside controlled environments.

- What intrinsic properties within datasets or model architectures influence the effectiveness and feasibility of adversarial attacks?
- What are the practical limitations that diminish the effectiveness of theoretically potent adversarial attacks in real-world settings?

Our paper is structured as follows: Sect. 2, “Background,” outlines different adversarial attacks. Section 3, “Related Work,” reviews developments and identifies research gaps. Section 4, “Methodology,” describes our experimental setup. Section 5, “Results,” displays our findings. In Sect. 6, “Discussion,” we interpret these results. Finally, Sect. 7, “Conclusion,” summarizes our key discoveries and their implications for future work.

## 2 Background

In machine learning, adversarial attacks are strategies designed to deceive algorithms by creating inputs to cause a model to make mistakes, known as adversarial examples. These attacks are particularly concerning because they can be unnoticeable to humans but lead to significant errors in machine learning systems. Adversarial attacks can occur during the training or testing phases [20]. They are broadly categorized into four types: poisoning attacks [18, 21], evasion attacks [20], membership inference attacks [22], and model inversion attacks [23]. Evasion and poisoning attacks are the main categories of attacks, with membership inference and model inversion attacks considered subcategories of them.

### *2.1 Poisoning Attacks*

Poisoning attacks occur during the training phase of a machine learning model. In this kind of attack, adversarial samples are crafted and injected into the training set of a machine learning system [21]. These attacks aim to disrupt the system's functionality or impair its decision-making capabilities. Poisoning attacks are primarily classified into two categories: data manipulation and label manipulation [5]. Label manipulation involves altering the labels of training data to disrupt the model's learning process, while data manipulation directly alters the features of training samples, such as embedding specific patterns or noises into them. We explored two types of poisoning attacks: the label-flipping attack [18] and the feature collision attack [19].

The label-flipping attack [18] is a label manipulation technique where labels in the training set are maliciously altered. The feature collision attack [19] is a data manipulation technique where the attacker creates an input similar to a legitimate input in terms of features but different enough to be classified into another category. By adding perturbations, the attack moves the feature-space representation of the base class closer to that of the target example [24].

**Table 1** Comparison of attack models

| Attack model | FGSM | BIM | JSMA | C&W | Transferability | Label flipping | Feature collision |
|---|---|---|---|---|---|---|---|
| Knowledge level | White-box | White-box | White-box | White-box | Black-box | White-box | White-box |
| Attack type | Untargeted | Targeted/ untargeted | Targeted/ untargeted | Targeted/ untargeted | Targeted/ untargeted | Targeted/ untargeted | Targeted |
| Model dependency | Model specific | Model specific | Model specific | Model specific | Model agnostic | Model agnostic | Model agnostic |
| Attack method | Gradient-based | Iterative gradient-based | Saliency map-based | Optimization-based | Transferability | Data poisoning | Data poisoning |
| Vulnerabilities | Model gradients | Model gradients | Feature sensitivity | Model confidence | Model generalization | Data integrity | |
| Potential impact | Misclassify | Misclassify | Misclassify | Misclassify | Misclassify | Model training corruption | Model training corruption |

## 2.2 *Evasion Attacks*

Evasion attacks occur during the testing phase of a machine learning model. Attackers create adversarial samples that appear normal to the model but have slight alternations designed to exploit its vulnerabilities. This can lead to significant security problems and reduce the model's performance, especially in systems where accurate detection is crucial. For evasion attacks, we utilized five different techniques: FGSM, BIM, JSMA, C&W, and transferability attacks.

FGSM [13] rapidly generates perturbations designed to maximize the model's loss by adding noise to the input in the direction of the gradient's sign [25]. BIM [14] is an iterative approach building on FGSM, iteratively updating the adversarial example. JSMA [15] manipulates features with the highest impact on classification by employing a saliency map derived from a Jacobian matrix to determine which input features most influence the model's classification [26]. The C&W attack [16] uses optimization techniques to find the smallest possible perturbation that causes the desired misclassification. The transferability attack [17] involves crafting adversarial examples for one machine learning model that can often mislead another distinct model, even if the two models have varied architectures.

Table 1 provides a comprehensive comparison of these adversarial attacks based on key dimensions. The knowledge level dimension differentiates between white-box attacks, which require complete knowledge of the model's architecture and parameters, and black-box attacks, which need minimal knowledge about the model. The attack type distinguishes between untargeted attacks, which aim to cause general misclassification, and targeted attacks, which are designed to misclassify inputs into specific incorrect classes. The model dependency indicates whether an attack is specific to certain types of models or can be applied universally. The attack method details the techniques used. Lastly, the vulnerabilities and potential impact sections examine the weaknesses these attacks exploit and the risks they pose.

# 3 Related Work

Debicha et al. [3] reviewed evasion attacks and defenses in the intrusion detection systems (IDS) domain, focusing on the feasibility in cybersecurity and including the NSL-KDD dataset—a refined version of KDD Cup 1999. Their review highlights the predominance of studies in controlled environments, underscoring the importance of realism and practical constraints in adversarial attack and defense development and evaluation. Their key contribution is an updated review of adversarial strategies, emphasizing their practicality and realism in IDS scenarios.

In another study, Venturi and Zanasi [4] provided an overview of poisoning and evasion attacks and countermeasures in cybersecurity, focusing on their practical applicability and feasibility. Their attacks are focused on the domain of malware detection and network intrusion detection systems. They analyze the feasibility and applicability of adversarial attacks in real-world scenarios for malware and network intrusion detection systems.

Apruzzese et al. [27] explored realistic poisoning and evasion attacks against ML-based network intrusion detection systems (ML-NIDS), emphasizing the practical side of cybersecurity. Their analysis covered studies on the KDD dataset, concluding that attacks manipulating real network traffic—where attackers have constrained feature knowledge and lack feedback or access to NIDS training data—are the most realistic. They stressed that the success and feasibility of attacks depend on the attacker's knowledge and the system's vulnerabilities in the NIDS domain.

Zhang et al. [28] delved into the feasibility of adversarial attacks in real-world scenarios, particularly within military and defense applications. They covered image classification, object detection, SAR systems, and speaker-recognition systems. They addressed the challenges of executing these attacks beyond controlled environments, emphasizing the military's unique constraints. They also offered insights into how varying degrees of system knowledge impact attack success rates.

In a different aspect of AI security, Tyukin et al. [6] introduced algorithms for stealth attacks on AI and deep learning systems, along with a theoretical analysis of their success, using the MNIST and CIFAR-10 datasets. Assuming control over the AI system for code or parameter modifications, they demonstrated the feasibility of such attacks, showing that altering a single neuron could significantly compromise AI systems. They concluded with mitigation recommendations through model design practices.

Similarly, Tramer et al. [29] investigated the vulnerabilities of ML models, centering on model extraction attacks. Their study reveals how adversaries could exploit publicly accessible query interfaces to steal ML models in the domain of ML-as-a-service, providing a crucial understanding of this specific threat vector. They showed that simple and computationally fast attacks could extract high-fidelity models.

Oliveira Junior et al. [30] explored the resilience of lexical sentiment analysis classifiers to adversarial attacks in social media, particularly by examining Twitter sentiment data. Their study shows that adversarial attacks can significantly distort classifiers' sentiment perception, misrepresenting actual social media sentiment. They highlight the importance of implementing defenses to protect sentiment analysis tools from such manipulations.

## 4 Methodology

### *4.1 Datasets*

To ensure a diverse range of data for our study, we carefully selected four distinct datasets, each presenting unique characteristics and challenges associated with various data types. Notably, much of the existing literature predominantly focuses on a single data type, such as image datasets, and often extends findings from these to all types of data. To address this limitation and provide a more comprehensive evaluation, we included a variety of datasets in our study.

The first dataset is the MNIST dataset, which stands for Modified National Institute of Standards and Technology. This is a dataset of handwritten digits, consisting of grayscale images, that is used in a huge number of ML studies [31–33]. Another dataset is the KDD Cup 1999 dataset, originating from the 1998 DARPA Intrusion Detection Evaluation Program, a widely utilized resource for intrusion detection algorithms in cybersecurity [34]. This dataset encompasses a variety of attack types and is classified into different unbalanced classes.

The MIT-BIH Arrhythmia Database, hosted by PhysioNet and created by the Massachusetts Institute of Technology and Beth Israel Hospital, comprises electrocardiogram (ECG) recordings instrumental in studying the heart's activity and diagnosing various cardiac conditions. This dataset is famous for machine learning studies [35, 36]. Lastly, we chose the Twitter sentiment dataset, a valuable resource in natural language processing and sentiment analysis designed to capture public opinion and emotions expressed through the Twitter platform. It is widely used in machine learning [37] and is categorized into three distinct sentiment classes: positive, negative, and neutral.

## *4.2 Machine Learning Models*

We evaluated the impact of adversarial attacks on five well-established models, each chosen for its unique architectural features and application domains. Decision trees (DTs)[7] are easy to interpret and visualize, making them suitable for simple tasks. Support vector machines (SVMs)[8] are particularly effective in high-dimensional spaces, making them well-suited for classification tasks where the number of features is high.

Convolutional neural networks (CNNs) [9] stand out in recognizing patterns in complex inputs through filtering relevant information. The architecture includes two convolutional layers with a kernel size of 5. This is followed by a dropout layer, a fully connected layer with an input size of 320, and a final fully connected layer with an input size of 50 and an output size of 10. The CNN is optimized using the Adam optimizer with a learning rate of 0.001 and a batch size of 64 and trained for five epochs.

Long short-term memory networks (LSTMs) [10], a type of recurrent neural network, excel with sequence data like text or time series. The LSTM model takes input embeddings of size 384 and uses an LSTM layer followed by a fully connected layer to output predictions across three classes. The model is trained using the Adam optimizer with a learning rate of 0.001 and a batch size of 64 and for five epochs.

Multilayer perceptrons (MLPs) [11] are composed of multiple layers of neurons used for tasks like classification and regression. The architecture consists of an input dimension followed by a hidden layer with 64 neurons, ReLU activation, a dropout layer with a rate of 0.5, and an output layer with neurons equal to the number of unique labels. The MLP is optimized using the Adam optimizer with a learning rate of 0.001 and a batch size of 64 and trained for five epochs.

These models, with distinct structures and uses, offer a broad perspective for assessing the impacts of adversarial attacks.

## 4.3 Data Preparation and Model Training

We standardized datasets to ensure compatibility with various models. MNIST images were normalized and reshaped for model input. The KDD dataset was encoded to handle its multiple features, revealing class imbalances. MIT-BIH ECG signals were filtered, normalized, and segmented for analysis. Twitter sentiment data was cleaned and transformed into embeddings.

We applied various models suited to the unique nature of each dataset. Decision trees and SVMs handled different data types well. CNNs were effective for patterns in the MNIST and MIT-BIH datasets. LSTM networks processed the sequential text data in the Twitter sentiment dataset. For the structured KDD dataset, we used MLP. Our training approach included dense and dropout layers in MLPs, as well as dropout with ReLU activations in CNNs. We optimized NN models using CrossEntropyLoss and the Adam optimizer, ensuring that each model was trained for maximum accuracy and adaptability to the specific dataset challenges.

Our evaluation measured model resilience after training using metrics like accuracy, precision, recall, and F1 score, which are crucial for imbalanced datasets. We also considered computational efficiency, loss metrics, MAE, MSE, RMSE, ROC-AUC scores, and confusion matrices. These metrics provided a detailed performance overview, emphasizing the impact of model design and resistance to adversarial attacks.

## 4.4 Adversarial Attack Implementation

FGSM, BIM, C&W, and JSMA are gradient-based attacks that need access to model gradients, making them ineffective for models like decision trees and SVMs, which do not provide gradient information. These attacks require complete knowledge of the model (white-box conditions), so we applied them only to neural networks, excluding decision trees and SVMs from these evaluations.

### 4.4.1 Evasion Attacks

We applied our evasion attacks using the Adversarial Robustness Toolbox (ART) [38]. For FGSM, we set up our PyTorch classifier and generated adversarial

**Table 2** Poisoning settings

| Poisoning parameter | MNIST | Twitter | KDD | MIT-BIH |
|---|---|---|---|---|
| Number of targets | 2000 | 2000 | 2000 | 2000 |
| Epsilon | 0.7 | 0.7 | 0.9 | 0.9 |
| Number of poisons per target | 50 | 50 | 100 | 60 |
| Poisoning ratio | 62% | 38% | 38% | 60% |

examples in one step using the loss gradient, adjusting inputs with an $\epsilon$ of 0.2 for optimal effect. We used BIM to assess model resilience, keeping $\epsilon$ at 0.2. BIM fine-tunes perturbations through feedback for precise adversarial example exploration. For JSMA, we target crucial input features with tailored theta (for perturbation magnitude) and gamma (for limiting the fraction of features adjusted) parameters to achieve minimal, effective changes. For the C&W attack, we minimize the L2 distance and misclassify inputs with precision. We also implemented transferability attacks in black-box settings by leveraging the FGSM method on a surrogate model. These examples aimed at misleading the surrogate were tested on the main model to assess if vulnerabilities transfer.

### 4.4.2 Poisoning Attacks

In our approach, the label-flipping attack alters 50% of the training data labels to incorrect classifications. For the feature collision attack, we introduced adversarial samples into the training data that share feature space with specific targets. This attack merges base class samples into the target instances' feature space to confuse the model. We selected 2000 target instances from each dataset and diversified them by incrementing class labels for base and target classes. Perturbations ($\epsilon$ between 0.7 and 0.9) generated 50–100 adversarial samples per target. These samples were integrated into the original data to create a poisoned dataset. Table 2 details our poisoning strategy.

### 4.4.3 Evaluating ML Model Resilience to Adversarial Attacks

After launching adversarial attacks, we evaluated their impact on target models using key metrics. We tracked the number of queries per attack, total prediction requests, and the interaction period, which is the timeframe in which the attacker actively interacts with the machine learning model. We also measured the runtime for each attack to assess their practicality and efficiency.

We then assessed model accuracy to determine the immediate impact on prediction capabilities and the attack success rate. We evaluated performance degradation to see how much the model's performance deteriorated due to the attacks. We recalculated precision, recall, and F1 scores to understand changes in sensitivity and specificity. Additionally, we quantified prediction errors with MAE, MSE, and RMSE. Confusion matrices provided a visual analysis of class performance, and ROC curves assessed the models' discriminatory capabilities post-attack. For feature collision attacks, we evaluated the targeted instance success rate to highlight the effectiveness of these strategies.

# 5 Experimental Results

## *5.1 Baseline Models' Performance*

Before evaluating our machine learning models' resilience to adversarial attacks, we established their baseline performance under normal conditions, as shown in Table 3. This initial evaluation highlights each classifier's strengths and accuracy across various datasets. The models demonstrated high accuracy on the MNIST, KDD, and MIT-BIH datasets, confirming their effectiveness in image classification, network intrusion detection, and ECG signal analysis, respectively. However, the performance on the Twitter sentiment dataset was lower, reflecting the complexity of sentiment analysis due to linguistic nuances, variability, and context ambiguity.

## *5.2 Models' Performance After Adversarial Attacks*

Assessing models' resilience to adversarial threats is crucial for determining an attack's success. An effective attack significantly reduces model performance, indicating a breach in model integrity. Table 4 shows that FGSM and BIM attacks have minimal impact on CNN models for MNIST and MIT-BIH, demonstrating CNN's robustness. However, attacks like JSMA and C&W drastically reduce performance in various cases, highlighting their general effectiveness. Notably, transferability attacks are particularly potent against DT and SVM classifiers, revealing their vulnerability to adversarial examples from neural networks, even without direct parameter access.

Table 5 demonstrates the impact of poisoning attacks. Label-flipping attacks were generally effective, but the KDD dataset showed resistance. Feature collision attacks, although not significantly affecting overall performance, had a noticeable

**Table 3** Performance of the models before adversarial attacks

| Dataset | DT | SVM | NN |
|---|---|---|---|
| MNIST | 87.76% | 97.92% | 98.88% |
| Twitter | 45.82% | 69.91% | 67.31% |
| KDD | 99.95% | 99.93% | 99.91% |
| MIT-BIH | 95.42% | 98.16% | 97.43% |

**Table 4** Performance after evasion attacks

| | Evasion attack | | | | Transferability attack | | |
|---|---|---|---|---|---|---|---|
| Dataset | FGSM | BIM | JSMA | C&W | DT | SVM | NN |
| MNIST | 98.16% | 98.00% | 0.26% | 78.92% | 4.21% | 2.17% | 98.21% |
| Twitter | 19.72% | 15.19% | 16.59% | 31.17% | 35.32% | 17.44% | 17.76% |
| KDD | 41.42% | 18.38% | 6.90% | 11.83% | 15.06% | 41.15% | 19.55% |
| MIT-BIH | 91.29% | 91.62% | 85.96% | 62.12% | 77.72% | 71.86% | 71.79% |

**Table 5** Performance of the models after poisoning attacks

| | Label flipping | | | Feature collision | | |
|---|---|---|---|---|---|---|
| | DT | SVM | NN | DT | SVM | NN |
| MNIST | 42.45% | 48.61% | 73.00% | 83.90% | 97.70% | 98.88% |
| Twitter | 36.42% | 45.16% | 35.96% | 45.20% | 70.19% | 61.80% |
| KDD | 72.86% | 99.28% | 99.52% | 99.95% | 99.38% | 99.83% |
| MIT-BIH | 47.28% | 43.44% | 77.28% | 91.03% | 93.44% | 91.17% |

**Table 6** Number of queries per attack

| Dataset | Test set size | FGSM | BIM | JSMA | C&W |
|---|---|---|---|---|---|
| MNIST | 10,000 | 20,000 | 110,000 | 7,778,170 | 6,961,651 |
| Twitter | 28,830 | 57,660 | 268,906 | 200,534 | 16,002,274 |
| KDD | 60,000 | 120,000 | 6,180,000 | 76,894,216 | 33,805,798 |
| MIT-BIH | 16,882 | 16,882 | 236,348 | 10,505,044 | 11,713,908 |

targeted impact on specific instances. Therefore, we will analyze these targeted impacts in greater detail in the next section.

## 5.3 *Number of Queries per Attack*

The computational demands of adversarial attacks, such as FGSM, BIM, JSMA, and C&W, vary in terms of the number of queries needed to generate adversarial examples. These differences arise due to the iterative nature of some methods. As shown in Table 6, transferability and poisoning attacks are more efficient, using a number of queries equivalent to the test set size to exploit vulnerabilities or measure impact with minimal interaction. This efficiency highlights the importance of implementing real-world defenses like rate-limiting, which can limit the number of queries and potentially block high-query attacks before they compromise the model. Therefore, for adversarial strategies to be effective in practice, it is crucial to minimize query numbers to avoid detection and bypass existing security measures, balancing attack effectiveness with interaction intensity.

**Table 7** Period of interaction for evasion attacks (measured in seconds)

| | Evasion attack | | | | Transferability attack | | |
|---|---|---|---|---|---|---|---|
| Dataset | FGSM | BIM | JSMA | C&W | DT | SVM | NN |
| MNIST | 6.34 | 507.65 | 9682.11 | 16299.20 | 4.35 | 121.44 | 8.87 |
| Twitter | 12.42 | 711.19 | 461.95 | 18726.32 | 0.03 | 1682.86 | 8.11 |
| KDD | 6.05 | 339.45 | 249360 | 35644.38 | 0.03 | 17.88 | 10.70 |
| MIT-BIH | 8.39 | 171.74 | 12755.25 | 15940.38 | 2.75 | 3.41 | 5.64 |

**Table 8** Period of interaction for poisoning attacks (measured in seconds)

| | Label flipping | | | Feature collision | | |
|---|---|---|---|---|---|---|
| Dataset | DT | SVM | NN | DT | SVM | NN |
| MNIST | 14.37 | 4365.00 | 257.85 | 230.16 | 10000.98 | 749.59 |
| Twitter | 138.51 | 13288.79 | 211.10 | 1135.33 | 22897.34 | 475.61 |
| KDD | 22.89 | 31743.34 | 72.23 | 13.53 | 4319.80 | 84.65 |
| MIT-BIH | 783.13 | 3992.89 | 243.42 | 2806.94 | 6296.84 | 164.70 |

## 5.4 *Period of Interaction*

The effectiveness of adversarial attacks depends heavily on their interaction time with target models, especially in real-world systems with strict time constraints. Complex techniques like JSMA and C&W attacks, as well as those targeting SVMs, may trigger defensive mechanisms due to their prolonged interaction times. These methods can struggle to be practical in environments with stringent activity monitoring and time restrictions. Our analysis, shown in Tables 7 and 8, demonstrates the impracticality of these methods against systems protected by time-based defenses, underlining the necessity for attacks to be both effective and time-efficient to ensure their real-world applicability and threat potential.

## 5.5 *Models' Performance After Feature Collision Attack*

Table 9 evaluates the success of feature collision attacks on 2000 targeted instances per dataset. The results varied, showing that attack success depends on data characteristics and target selection, not just model type. For example, the same CNN model had different success rates on MNIST and MIT-BIH. The data suggest that SVMs are robust against feature collision attacks, especially on the KDD dataset, while decision trees are more vulnerable. In comparison, label-flipping attacks were consistently effective across models, making them a more reliable poisoning method.

**Table 9** Percentage of target instances successfully attacked in feature collision attack

| Dataset | DT | SVM | NN |
|---|---|---|---|
| MNIST | 20.70% | 0.35% | 0.85% |
| Twitter | 54.85% | 1.85% | 29.85% |
| KDD | 0.15% | 0.80% | 99.15% |
| MIT-BIH | 16.85% | 14.50% | 20.25% |

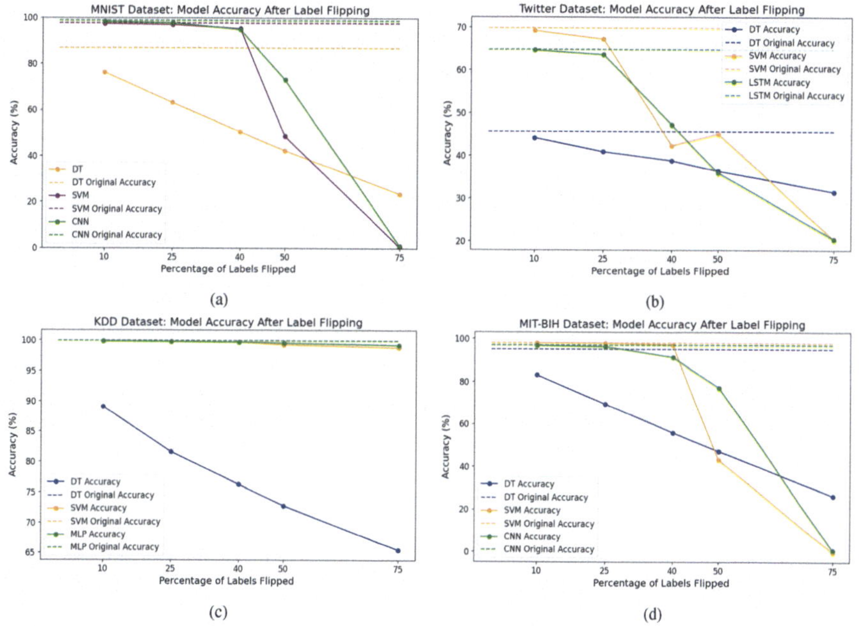


**Fig. 1** Label-flipping effect on different datasets. (**a**) MNIST dataset. (**b**) Twitter dataset. (**c**) KDD dataset. (**d**) MIT-BIH dataset

## 5.6 *Evaluating Various Poisoning Ratios in Label-Flipping Attacks*

In Sect. 5.2, we initially reported the effects of flipping 50% of the labels, which is an unrealistic scenario. In real-world situations, such extensive manipulation is improbable, as attackers typically have limited access and large-scale alterations would be easily detectable. To reflect more realistic conditions, we investigated smaller proportions of label flipping. Our results, shown in Fig. 1, reveal varying levels of accuracy impact on models when subjected to different percentages of label flipping across datasets. SVM and NN models remain robust against minor label changes, experiencing significant performance drops only with extensive data alteration. In contrast, DT classifiers are highly sensitive, showing notable performance degradation even with minimal label flipping. This indicates that in real-world scenarios with low proportions of label flipping, only weaker models exhibit significant performance declines to this attack, while other models demonstrate greater resilience.

### 5.7 *Impact of Varying Levels of Perturbation on Attack Success*

The perturbation level, controlled by the epsilon parameter, creates a trade-off between attack strength and detectability. Higher epsilon values make the attack more effective but also more noticeable and easier to detect. Attackers aim to keep epsilon low for undetectable attacks while still compromising model performance. In Fig. 2, we see how varying epsilon levels affect different datasets' accuracy. For image classification and pattern recognition tasks (Fig. 2a and d), accuracy steadily declines as epsilon increases, indicating these models' vulnerability to even minor perturbations. The KDD dataset shows a sharp initial accuracy drop, then a plateau, suggesting resilience up to a certain perturbation threshold. The Twitter dataset's sentiment analysis models remain stable with increasing epsilon, likely due to natural language processing models' inherent resistance to noise. Thus, while

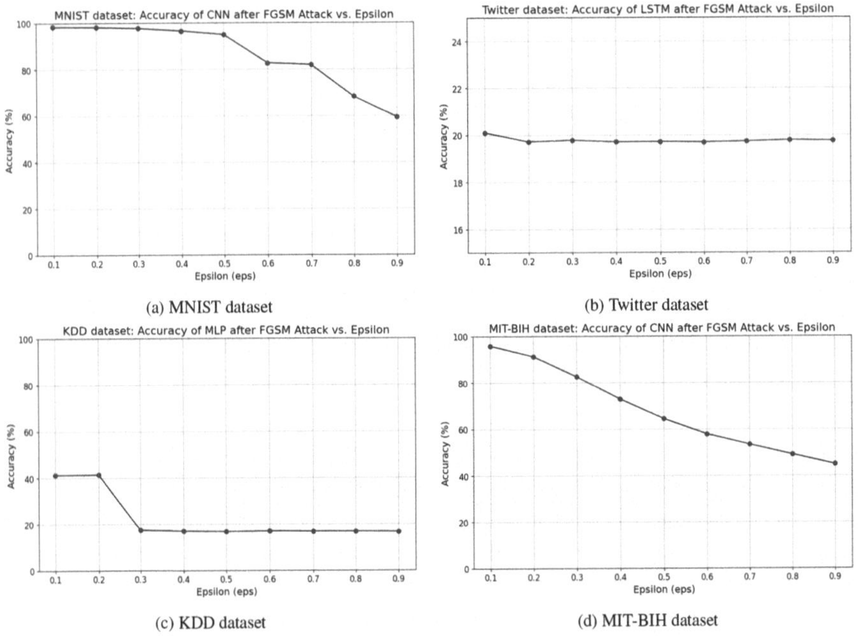


(a) MNIST dataset (b) Twitter dataset (c) KDD dataset (d) MIT-BIH dataset

**Fig. 2** Impact of varying FGSM epsilon values on model accuracy. (**a**) MNIST dataset. (**b**) Twitter dataset. (**c**) KDD dataset. (**d**) MIT-BIH dataset higher perturbations enhance attack effectiveness, they also risk detection. Keeping perturbation levels low makes real-world attacks less effective but harder to detect.

### *5.8 Influence of Attacker's Knowledge on Adversarial Attack Success*

In adversarial machine learning, the attacker's knowledge level is critical for attack success, as shown in Table 10, based on our experiments. This table illustrates the level of knowledge or access required for each attack type to be successful. It means that to perform each attack, the attacker needs a specific amount of knowledge or access to execute it effectively. The table shows that having limited access or knowledge makes the attack unsuccessful and unperformable. Evasion attacks like FGSM, BIM, JSMA, C&W, and transferability need no dataset knowledge but vary in model knowledge requirements. Poisoning attacks like label flipping and feature collision depend on dataset access. Evasion attacks usually require detailed knowledge of the model (white box), except for transferability attacks, which work without detailed model knowledge (black box) and pose widespread risks. Feature collision attacks require knowledge of both the model and the dataset. This analysis shows that while white-box attacks are powerful, their high knowledge requirements limit real-world applicability. In contrast, black-box attacks are more feasible due to their lower dependency on specific model or dataset knowledge. Limited access or knowledge significantly reduces the effectiveness and feasibility of these attacks.

## 6 Discussion

In our comprehensive evaluation of machine learning model resilience across various datasets and attack types, we identified distinct patterns of vulnerability and robustness. We analyzed datasets such as MNIST, Twitter sentiment, KDD, and MIT-BIH to understand how different models withstand adversarial attacks. First, we will discuss the effects of adversarial attacks on each dataset separately.

For the MNIST dataset, we found it resilient against FGSM, BIM, and feature collision attacks, showing that minor perturbations are not enough to compromise the data significantly. However, C&W, JSMA, and label-flipping attacks were highly effective, severely impacting model performance. Additionally, simple adversarial examples generated by neural networks could disrupt models like decision trees and SVMs through transferability attacks.

The Twitter sentiment dataset showed consistent vulnerability across various attacks, regardless of their complexity. This could be due to the nuances of sentiment analysis, where even slight changes can lead to misclassifications, making all attack types equally harmful.

**Table 10** Level of attacker's knowledge across different attack models

| Level of attacker's knowledge | FGSM | BIM | JSMA | C&W | Transferability | Label flipping | Feature collision |
|---|---|---|---|---|---|---|---|
| Level of knowledge of the model | | | | | | | |
| 1. No knowledge | X | X | X | X | ✓ | ✓ | X |
| 2. Partial knowledge-model type | X | X | X | X | ✓ | ✓ | X |
| 3. Partial knowledge-model family | X | X | X | X | ✓ | ✓ | X |
| 4. Partial knowledge-model architecture | X | X | X | X | ✓ | ✓ | X |
| 5. Considerable knowledge-parameters and weights | ✓ | ✓ | ✓ | ✓ | ✓ | ✓ | ✓ |
| 6. Full knowledge | ✓ | ✓ | ✓ | ✓ | ✓ | ✓ | ✓ |
| Level of knowledge of the dataset | | | | | | | |
| 1. No knowledge | ✓ | ✓ | ✓ | ✓ | ✓ | X | X |
| 2. Dataset type | ✓ | ✓ | ✓ | ✓ | ✓ | X | X |
| 3. Feature knowledge | ✓ | ✓ | ✓ | ✓ | ✓ | X | X |
| 4. Statistical knowledge | ✓ | ✓ | ✓ | ✓ | ✓ | X | ✓ |
| 5. Sample knowledge | ✓ | ✓ | ✓ | ✓ | ✓ | X | ✓ |
| 6. Full dataset access | ✓ | ✓ | ✓ | ✓ | ✓ | X | ✓ |
| 7. Partial dataset access | ✓ | ✓ | ✓ | ✓ | ✓ | ✓ | ✓ |
| 8. Full dataset access | ✓ | ✓ | ✓ | ✓ | ✓ | ✓ | ✓ |

In the KDD dataset, a structured tabular format, evasion attacks were highly effective, significantly degrading performance. However, the dataset resisted poisoning attacks, likely because the structured nature of tabular data helps mitigate the effects of label manipulations and feature collisions. Evasion attacks, which exploit decision boundaries, were more damaging.

For the MIT-BIH dataset, we found resilience against FGSM, BIM, C&W, and JSMA attacks. However, simpler models like SVMs and decision trees were vulnerable to straightforward adversarial examples created using neural networks. Additionally, the MIT-BIH dataset was susceptible to poisoning attacks, indicating that malicious training manipulations could significantly impact cardiac signal data.

Then, we discuss the characteristics and implications of different attack types. Among evasion attacks, FGSM and BIM stand out for their speed in generating adversarial examples, but their impact is usually less severe than that of C&W and JSMA. JSMA and C&W attacks are more effective but require significantly more time due to their targeted attack nature, aiming for specific malicious outcomes at a computational time trade-off. A limitation of FGSM, BIM, C&W, and JSMA is their reliance on gradient-based methods, making them ineffective for models without gradient information, like DT and SVM. This reliance also limits their real-world applicability due to the need for a white-box setting.

Transferability attacks use NNs to generate adversarial examples that are effective even when applied to other types of models. This shows that NNs can create strong adversarial examples that can fool different models, including traditional ones like DTs and SVMs. Transferability attacks are particularly significant because they operate as black-box attacks, making them one of the closest types of attacks to real-world scenarios where attackers typically have limited access to model details.

Label-flipping attacks were effective across several datasets but less so on the KDD dataset due to its unique class distributions. Minor training alterations like label flipping were insufficient to cause misclassifications in KDD. Feature collision attacks, being targeted, did not significantly alter overall accuracy but were notably effective on the Twitter dataset. KDD and MNIST datasets remained robust against feature collision attacks, while MIT-BIH showed moderate vulnerability. NNs were significantly affected by these attacks, whereas DTs and SVMs displayed greater resilience.

## *6.1 Real-World Feasibility of Adversarial Attacks*

In controlled environments, adversarial attacks seem simple because they assume near-perfect information about the data and access the attacker has, which are unrealistic in real-world scenarios. Table 11 represents the likelihood of an attack occurring based on factors like complexity, knowledge, accessibility, resources, target vulnerabilities, query numbers, interaction time, and detectability.

The complexity of an attack makes it less practical in the real world. Attacks requiring detailed information about the model or data are less feasible due to

**Table 11** Likelihood of occurrence of attacks

| | FGSM | BIM | JSMA | C&W | Transferability | Label flipping | Feature collision |
|---|---|---|---|---|---|---|---|
| Complexity | Low | Medium | High | High | Medium | Low | Medium |
| Knowledge | High | High | High | High | Low/med | Low | Med/high |
| Accessibility | High | High | High | High | Low | High | High |
| Resource | Low | Low | High | High | Medium | Low | Medium |
| Vulnerabilities | High | High | Low | Low | Medium | High | Medium |
| Queries | Low | Med | High | High | Low | Low | Low |
| Interaction | Low | Low | High | High | Low | Med | Med |
| Detectability | High | Med | Low | Low | Medium | High | Low |
| Overall likelihood | Med | Med | Low | Low | High | Med | Med |

limited real-world access. Accessibility is another issue; many attacks require access to the model or data, which is often restricted. High computational demands limit the feasibility of many attacks for those without significant resources. Target vulnerabilities are also critical; common vulnerabilities are well-guarded, so attackers must find less-known and more complex vulnerabilities. The number of queries needed for an attack can trigger defensive mechanisms, blocking the attack. Limited interaction time due to security measures can prevent prolonged attacks. Detectability is another factor; easily detectable attacks are less effective as they can be quickly countered.

Overall, each attack's likelihood of occurrence varies based on these factors. Among the attacks discussed, only black-box attacks have a high chance of being successful in real-world scenarios. Due to these practical limitations, many attacks evaluated in the literature have a low likelihood of occurrence outside controlled environments.

# 7 Conclusion

This study examined the practical application of adversarial attacks on machine learning models, comparing theoretical vulnerabilities with real-world scenarios. We analyzed various datasets and models to understand how machine learning systems resist different adversarial strategies, including evasion and poisoning attacks. Unlike previous studies that focused on specific data types or models, our research aimed to generalize the impact of various adversarial attacks, providing a broader perspective on their effectiveness and feasibility.

A key objective was to identify the intrinsic properties of datasets and model architectures that influence the success and feasibility of adversarial attacks. Our findings suggest that focusing solely on specific data or models overlooks the diverse vulnerabilities across different machine learning systems. By using a wide range of models and datasets, we could draw general conclusions about the factors that significantly affect attack feasibility, highlighting the importance of diversity in assessing machine learning security.

Furthermore, we aimed to evaluate the feasibility of deploying adversarial attacks on machine learning models in real-world settings compared to controlled environments. Our investigation highlighted several factors, often overlooked in the literature, that significantly influence the practical execution of such attacks. These factors reduce the likelihood of successful real-world adversarial attacks, suggesting that the threat of many theoretically strong attacks is less severe when considering operational constraints and common defense strategies. Our analysis shows that most white-box attacks, requiring extensive knowledge of the target model or dataset, are limited in practical application. This finding offers reassurance to stakeholders in machine learning applications.

## 8 Future Work

**Broader Model and Dataset Coverage** Future research should aim to expand the range of models and datasets examined in adversarial attack studies, especially those relevant to specialized or high-stakes applications like healthcare, finance, and critical infrastructure. Expanding the scope in this way would improve the generalizability of the findings and provide a more comprehensive understanding of how different models and datasets respond to adversarial threats.

**Defense Mechanisms and Emerging Paradigms** We also recommend prioritizing the development and evaluation of defense mechanisms against more plausible black-box attacks, which are more reflective of real-world scenarios. Strengthening defenses in this area would significantly enhance the security of machine learning models against the most likely threats. Additionally, exploring the feasibility of adversarial attacks within emerging machine learning paradigms could yield valuable insights into preemptive security measures, thereby ensuring the resilience of future machine learning systems against evolving adversarial threats.

## References


1. Q. An, S. Rahman, J. Zhou, J.J. Kang, A comprehensive review on machine learning in healthcare industry: classification, restrictions, opportunities and challenges. Sensors **23**(9), 4178 (2023)
2. J. Stilgoe, Machine learning, social learning and the governance of self-driving cars. Soc. Stud. Sci. **48**(1), 25–56 (2018)
3. I. Debicha, B. Cocheza, T. Kenazac, T. Debattyb, J. Dricota, W. Meesb, Review on the feasibility of adversarial evasion attacks and defenses for network intrusion detection systems. Inform. Secur. Appl. (2023)
4. A. Venturi, C. Zanasi, On the feasibility of adversarial machine learning in malware and network intrusion detection, in *2021 IEEE 20th International Symposium on Network Computing and Applications (NCA)* (2021)
5. Z. Tian, L. Cui, J. Liang, S. Yu, A comprehensive survey on poisoning attacks and countermeasures in machine learning. ACM Comput. Surv. **55**(8), 166 (2022)

6. I.Y. Tyukin, D.J. Higham, A. Bastounis, E. Woldegeorgis, A.N. Gorban, The feasibility and inevitability of stealth attacks (2023). arXiv:2106.13997v4
7. J.R. Quinlan, *Induction of Decision Trees*. Machine Learning, vol. 1 (Kluwer Academic Publishers, Boston, 1986), pp. 81–106
8. R.G. Brereton, G.R. Lloyd, Support vector machines for classification and regression. Roy. Soc. Chem. (2009). https://doi.org/10.1039/b918972f
9. Y. LeCun, B. Boser, J.S. Denker, D. Henderson, R.E. Howard, W. Hubbard, L.D. Jackel, Backpropagation applied to handwritten zip code recognition. IEEE **1**(4), 541–551 (1989)
10. S. Hochreiter, J. Schmidhuber, Long short-term memory, in *Neural Computation* (1997)
11. D.E. Rumelhart, G.E. Hinton, R.J. Williams, *Learning Representations by Back-Propagating Errors* (Nature Publishing Group, 1986)
12. Y. Lecun, L. Bottou, Y. Bengio, P. Haffner, Gradient-based learning applied to document recognition. IEEE **86**(11), 2278–2324 (1998)
13. I.J. Goodfellow, J. Shlens, C. Szegedy, Explaining and harnessing adversarial examples, in *ICLR* (2015)
14. A. Kurakin, I.J. Goodfellow, S. Bengio, Adversarial examples in the physical world, in *ICLR* (2017)
15. N. Papernot, P. McDaniel, S. Jha, M. Fredrikson, Z.B. Celik, A. Swami, The limitations of deep learning in adversarial settings (2015). arXiv:1511.07528v1
16. N. Carlini, D. Wagner, Towards evaluating the robustness of neural networks (2017). arXiv:1608.04644v2
17. N. Papernot, P. McDaniel, Transferability in machine learning: from phenomena to black-box attacks using adversarial samples (2016). arXiv:1605.07277v1
18. B. Biggio, P. d'Armi, B. Nelson, P. Laskov, Poisoning attacks against support vector machines (2013). arXiv:1206.6389v3.
19. A. Shafahi, W.R. Huang, M. Najibi, O. Suciu, C. Studer, T. Dumitras, T. Goldstein, Poison frogs! targeted clean-label poisoning attacks on neural networks (2018). arXiv:1804.00792v2
20. Q. Liu, P. Li, W. Zhao, W. Cai, S. Yu, V.C.M. Leung, A survey on security threats and defensive techniques of machine learning: a data driven view. Access **6**, 12103–12117 (2018)
21. B. Biggio, B. Nelson, P. Laskov, Support vector machines under adversarial label noise, in *Asian Conference on Machine Learning* (PMLR, 2011), pp. 97–112
22. R. Shokri, M. Stronati, C. Song, V. Shmatikov, Membership inference attacks against machine learning models (2017). arXiv:1610.05820v2
23. M. Fredrikson, S. Jha, T. Ristenpart, Model inversion attacks that exploit confidence information and basic countermeasures. ACM (2015). https://doi.org/10.1145/2810103.2813677
24. M. Goldblum, D. Tsipras, C. Xie, X. Chen, A. Schwarzschild, D. Song, A. Madry, B. Li, T. Goldstein, Data security for machine learning: data poisoning, backdoor attacks, and defenses (2020). arXiv:2012.10544
25. K. Ren, T. Zheng, Z. Qin, X. Liu, *Adversarial Attacks and Defenses in Deep Learning* (Elsevier, Amsterdam, 2020)
26. A.H. Qureshi, H. Larijani, M. M. Yousefi, A. Adeel, N. Mtetwa, An adversarial approach for intrusion detection systems using jacobian saliency map attacks (JSMA) algorithm. Computers **9**(3), 58 (2020)
27. G. Apruzzese, M. Andreolini, L. Ferretti, M. Marchetti, M. Colajanni, Modeling realistic adversarial attacks against network intrusion detection systems. Digit. Threat. Res. Pract. **3**(3), 31 (2022)
28. L.A. Zhang, G.S. Hartnett, J. Aguirre, A.J. Lohn, I. Khan, M. Herron, C. O'Connell, *Operational Feasibility of Adversarial Attacks Against Artificial Intelligence* (RAND Corporation, 2022)
29. F. Tramer, F. Zhang, A. Juels, M.K. Reiter, T. Ristenpart, Stealing machine learning models via prediction APIs (2016). arXiv:1609.02943v2
30. G.A. de Oliveira Júnior, R.T. de Sousa Jr., R.O. Albuquerque, L.J.G. Villalba, Adversarial attacks on a lexical sentiment analysis classifier. Science Direct Comput. Commun. (2021). https://doi.org/10.1016/j.comcom.2020.08.015

31. A. Uprety, D.B. Rawat, Mitigating poisoning attack in federated learning, in *IEEE Symposium Series on Computational Intelligence (SSCI)* (2021). https://doi.org/10.1109/SSCI50451.2021.9659839
32. S. Wolf, W. Gamboa, M. Borowczak, Jangseung: a guardian for machine learning algorithms to protect against poisoning attacks, in *IEEE International Smart Cities Conference (ISC2)* (2021)
33. X. Yang, X. Qian, R. Zhang, N. Huang, H. Xia, *Low-Poisoning Rate Invisible Backdoor Attack Based on Important Neurons* (Springer Nature Switzerland AG, 2022)
34. E. Ashraf, N.F.F. Areed, H. Salem, E.H. Abdelhay, A. Farouk, FIDChain: federated intrusion detection system for blockchain-enabled IoT healthcare applications. Healthcare (2022). https://doi.org/10.3390/healthcare10061110
35. V. Singh, S. Tewary, V. Sardana, H.K. Sardana, Arrhythmia detection—a machine learning based comparative analysis with MIT-BIH ECG data, in *5th International Conference for Convergence in Technology (I2CT)*, Pune, India (2019)
36. S. Torres-Alegre, J. Fombellida, J. Pinuela-Izquierdo, D. Andina, AMSOM: artificial metaplasticity in SOM neural networks-application to MIT-BIH arrhythmias database, in *5th International Work-Conference on the Interplay Between Natural and Artificial Computation (IWINAC)* (2020)
37. G. Wang, T. Wang, H. Zheng, B.Y. Zhao, Man vs. machine: practical adversarial detection of malicious crowdsourcing workers, in *23rd USENIX Security Symposium* (2014)
38. M.-I. Nicolae, M. Sinn, M.N. Tran, B. Buesser, A. Rawat, M. Wistuba, V. Zantedeschi, N. Baracaldo, B. Chen, H. Ludwig, I.M. Molloy, B. Edwards, Adversarial robustness toolbox v1.0.0 (2019). arXiv:1807.01069v4

# Enhanced Public Safety: Real-Time Crime Detection with CNN-LSTM in Video Surveillance

**Chandana Thirunagari and Lilatul Ferdouse**

## 1 Introduction

In recent years, the integration of machine learning algorithms with video surveillance systems has shown immense potential in enhancing public safety. Video surveillance is a critical tool for monitoring public spaces, deterring criminal activities, and providing valuable evidence for investigations. As urban areas expand and the population grows, the volume of surveillance footage has increased exponentially, making manual monitoring impractical and inefficient. The need for automated systems that can analyze vast amounts of data in real time has become increasingly urgent.

Public safety is further bolstered by the deployment of Internet of Things (IoT) devices, such as smart cameras, motion detectors, and environmental sensors, which continuously gather data from various sources. These devices work together to create a networked surveillance infrastructure that can provide comprehensive coverage of public spaces. However, the sheer volume and diversity of data generated by these IoT devices present significant challenges. Traditional methods often struggle to process this data effectively, leading to delays and inaccuracies in detecting potential threats.

To address these challenges, advanced machine learning (ML) techniques are essential. ML algorithms can efficiently process large datasets, recognize complex patterns, and make accurate predictions in real time. By leveraging the power of ML, this work aims to develop a crime detection system that enhances the capabilities of video surveillance networks. The system focuses on the real-time detection

C. Thirunagari · L. Ferdouse (✉)
Physics and Computer Science, Wilfrid Laurier University, Waterloo, ON, Canada
e-mail: thir5380@mylaurier.ca; lferdouse@wlu.ca



I. Woungang, S. K. Dhurandher (eds.), *The 7th International Conference on Wireless, Intelligent and Distributed Environment for Communication*,
Lecture Notes on Data Engineering and Communications Technologies 237,
https://doi.org/10.1007/978-3-031-80817-3_3

and classification of criminal activities, providing timely alerts to law enforcement agencies. This integration of ML with video surveillance represents a significant step forward in improving public safety and ensuring that cities remain safe and secure environments for their inhabitants.

## 2 Video Surveillance and Deep Learning Techniques

The implementation of video surveillance and predictive analytics in enhancing public safety is a multifaceted approach that has been extensively researched. This section presents various methodologies, prediction models, and effective strategies that are used for public safety framework in smart cities. These studies provide valuable insights into the integration of advanced technologies such as deep learning, machine learning, and IoT in public safety systems. By examining different approaches to video analytics, crime prediction, and resource allocation, we can identify the most effective strategies for developing a robust public safety framework in smart cities.

The authors in [1] focus on utilizing deep learning method in video surveillance for analyzing crowded environments. Video data from public events is gathered and pre-processed by extracting frames and applying data augmentation techniques. CNNs and recurrent neural networks (RNNs) are used to detect and track individuals in crowds, with TensorFlow facilitating model implementation. The authors evaluate the models' accuracy in identifying abnormal behaviors and security threats, showing that deep learning significantly enhances crowd analysis capabilities, with accuracy rates reaching up to 88%.

An efficient violence detection method is proposed in [2]. The authors introduce a novel approach for detecting violent events in video surveillance footage. The proposed model integrates a U-Net-like network with Mobile-Net V2 as the encoder to extract spatial features and utilizes LSTM layers for capturing temporal information. This architecture is designed to be computationally efficient while maintaining high performance. The model achieved an average accuracy of 82% and average precision of 81% on the RWF-2000 dataset, which consists of complex real-world security camera footage.

In [3], the authors propose a novel approach for detecting and eliminating violent scenes in movies using a three-step process. The method segments movies into shots, selects representative frames based on saliency, and classifies them using a lightweight deep learning model fine-tuned with transfer learning. The proposed model is evaluated on three benchmark datasets and demonstrates superior performance, achieving up to 99.5% accuracy. This method outperforms existing techniques, including bag-of-words, support vector machines, and random forests, by providing faster and more accurate violence detection. The approach also shows versatility in different domains like surveillance and movies. Similarly, the authors in [4] propose a robust IoT-enabled anomaly detection system for smart city surveillance. The paper introduces a hybrid model combining 2D convolutional neural networks (CNNs) and echo state networks (ESN) for efficient anomaly detection in large-scale video data. The CNN extracts features from video frames, which are then refined by an auto-encoder and analyzed by the ESN for sequence learning and anomaly detection. This lightweight model is designed for implementation on edge devices, making it suitable for real-time applications in AIoT environments. The proposed approach outperforms traditional methods in detecting anomalies, demonstrating improved performance on challenging surveillance datasets.

### *2.1 Analysis of Existing Work*

The study on crowd analysis in [1] uses CNNs and RNNs, demonstrating significant improvements in detecting abnormal behaviors with an accuracy of up to 88%. This highlights the power of deep learning in enhancing the monitoring of crowded environments. However, the reliance on pre-collected data may restrict the model's adaptability to different real-world scenarios.

Violence detection approach in [2] employs a hybrid model combining U-Net-like networks and LSTM layers, achieving 82% accuracy and 81% precision. This method's computational efficiency is notable, though its effectiveness might be limited by the complexity of real-world footage and the need for diverse datasets to improve robustness. On the other hand, the authors in [3] propose a novel method for eliminating violent scenes from movies using a deep learning model fine-tuned with transfer learning, reaching up to 99.5% accuracy. This impressive performance underscores the potential of advanced deep learning techniques but also points to the need for broad dataset coverage to ensure the model's generalizability across different contexts. The IoT-enabled anomaly detection system integrates 2D CNNs and echo state networks (ESNs), achieving improved real-time performance with better accuracy in anomaly detection. However, its dependence on IoT infrastructure might limit its implementation in areas lacking advanced technological support. Enhancing data diversity and addressing infrastructure challenges are crucial for advancing these methodologies.

## 3 Proposed Integration of CNN with LSTM Model

The analysis of existing work indicates that CNN and LSTM are among the most popular machine learning models employed in video surveillance, each with unique benefits and limitations. This work primarily aims to develop a robust crime detection system that utilizes the capabilities of CNN and LSTM models for processing video surveillance data. The proposed CNN-LSTM system is designed to enhance public safety by detecting and predicting criminal activities in real time. This will be accomplished by leveraging advanced machine learning algorithms, with CNNs responsible for spatial feature extraction and LSTM networks handling temporal sequence modeling. In the following section, we will first outline the necessary data processing methods required to convert video data into fixed-sized

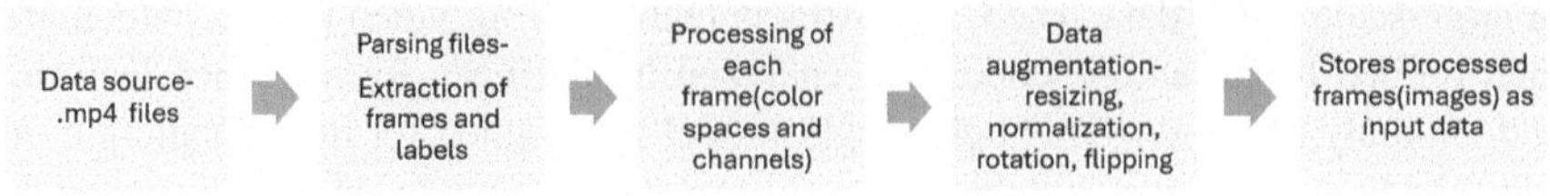


**Fig. 1** Video surveillance data pre-processing

frames. Then, we will describe a standalone CNN model, followed by the integrated CNN-LSTM model.

As shown in Fig. 1, the below steps are followed for data pre-processing:

*Data Source* The dataset [5] consists of videos categorized into seven classes: abuse, arrest, assault, burglary, fighting, robbery, and normal activities. Each class contains multiple MP4 files.

*Frame Extraction* Frames are extracted from each video at regular intervals to balance the computational load while preserving essential temporal information. This process converts video data into a series of images for further processing [6].

*Data Augmentation* To improve the model's robustness and generalization capability, the following data augmentation techniques are applied:

- *Random Horizontal Flipping*: This technique flips the images horizontally with a 50% chance. It helps the model learn features that are invariant to the horizontal orientation of the objects.
- *Random Rotation*: This technique rotates the images randomly within a specified range. It allows the model to be more robust in terms of the variations in object orientation.
- *Resizing*: All images are resized to a uniform size (224×224 pixels) to ensure consistency in input dimensions across the dataset [7].
- *Normalization*: Pixel values are normalized to a range of 0–1 by dividing by 255. This helps in faster convergence during training.

*Transformation Pipeline* Data pre-processing step is implemented as transformation pipeline that ensures that the input data is in the required format for deep learning models. This includes resizing frames to 224×224 pixels, normalizing pixel values, and converting images to tensors ( i.e., n-dimensional array).

## *3.1 CNN Model (Model 1)*

Following the necessary data pre-processing steps, a deep learning model is employed to analyze and predict outcomes. Figure 2 illustrates the CNN architecture utilized for video surveillance data analysis, comprising the following layers:

**Input Layer** Frames of size 224×224 pixels are fed into the CNN.

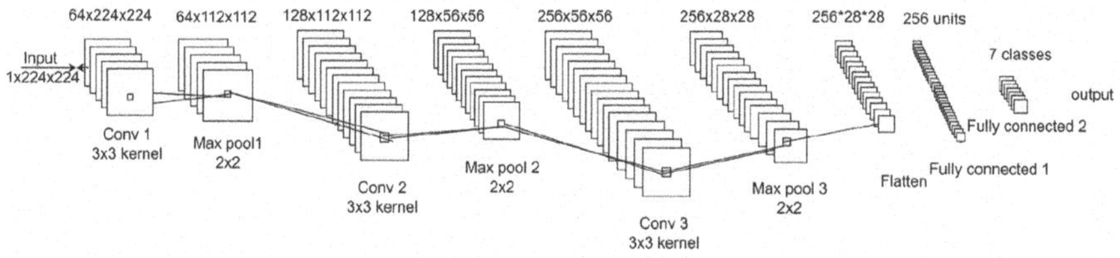


**Fig. 2** Model 1 ( CNN) architecture for video surveillance data processing

**Convolutional Layers** The network consists of multiple convolutional layers to extract spatial features from the input frames. *Convolutional layer* applies a set of filters (or kernels) to the input image to produce feature maps. Each filter detects different features like edges, textures, etc. *ReLU activation function* is used after each convolutional layer to introduce non-linearity. It replaces negative values with zero, which helps in mitigating the vanishing gradient problem [8]. Each convolution layer is followed by *max-pooling* layer, which down-samples the feature maps by taking the maximum value from each region of the feature map. This reduces the spatial dimensions and helps in reducing computational load and over-fitting.

**Flattening** The output from the final convolutional layer is flattened to form a 1D feature vector [9].

**Fully Connected Layers** Two fully connected layers are used to process the flattened features.

**Output Layer** The final layer has seven output nodes corresponding to the seven classes, with a softmax activation function to produce probability scores.

## *3.2 Integration of CNN and LSTM Model (Model 2)*

As shown in Fig. 3, the hybrid model consists of CNN and LSTM, which consists of the following layer:

**Input Layer** Frames of size 224×224 pixels are fed into the CNN.

**Convolutional Layers** Similar to Model 1, convolutional layers extract spatial features from the frames.

*Convolutional Layer* Three convolutional layers apply a set of filters to the input image to produce feature maps.

*LeakyReLU Activation Function* Instead of using ReLU, here LeakyReLU is used after convolution layer. Similar to ReLU, LeakyReLU allows a small, nonzero gradient when the unit is not active, which helps in learning from negative values. The LeakyReLU rate is set to 0.01 [8].

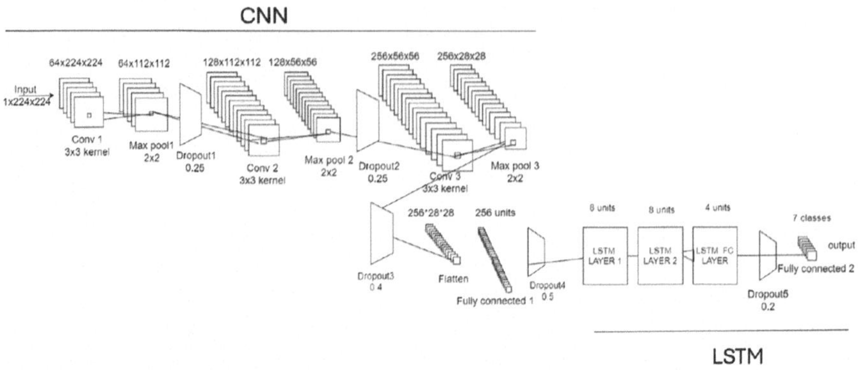


**Fig. 3** Model 2 (Hybrid) architecture for video surveillance data processing

*Max-Pooling Layer* Down-samples the feature maps with a pooling size of $2\times2$.

*Dropout Layers* In order to prevent the overfitting problem, after each max-pooling layer, a dropout layer is added. Dropout layer helps to drop out the unnecessary feature values in each frame. According to [10], dropout layers with dropout rate is set to 0.25 after the first two max-pooling layers, 0.4 after the third max-pooling layer, and 0.5 after the flattening step.

**Flattening** The output is flattened to form a 1D feature vector and passed to an LSTM module.

**LSTM Module** An LSTM module is integrated between the flattening and fully connected layers in the CNN model. The purpose of adding the LSTM module is to capture the temporal dependencies among the 1D feature values. The LSTM module comprises two LSTM layers and dropout layers. The output feature vectors from the CNN are fed into an LSTM network to capture temporal dependencies between frames. The first LSTM layer reduces the feature dimensions from 1 to 8. The second LSTM layer processes these eight-dimensional features [11]. Also a dropout rate of 0.2 is applied between LSTM layers to prevent overfitting problem.

**Fully Connected Layer** A final fully connected layer processes the output from the LSTM layers to produce class probabilities.

## 4 Performance Evaluations

### *4.1 Exploratory Data Analysis (EDA) and Simulation Environment*

**Class Distribution** Analyzing the distribution of video frames across different classes to ensure a balanced dataset.

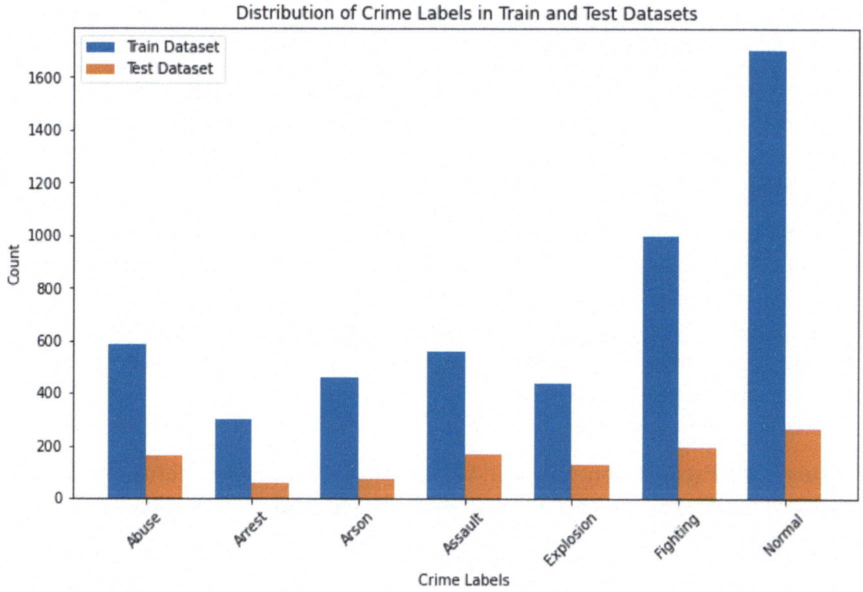


**Fig. 4** Distribution of train data

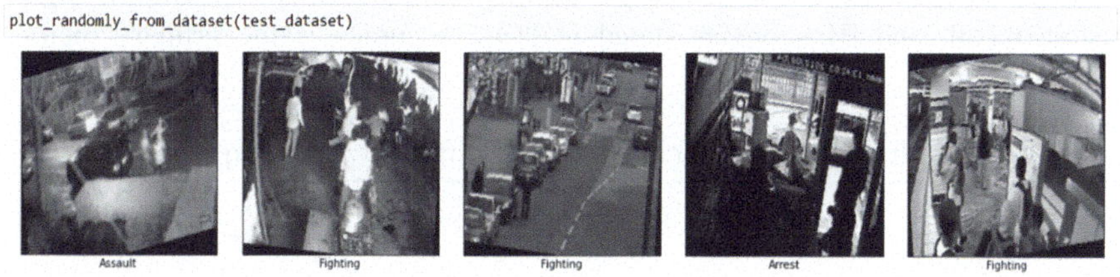


**Fig. 5** Random image extracted from videos

Figure 4 shows the distribution of train and test data of seven classes: abuse, arrest, arson, assault, explosion, fighting, and normal. This visual representation helps in understanding the frequency of different types of crime labels within the dataset. In train data, the normal class has almost 1600 images, while the remaining range from 600 to 1000. However, in test data, the distribution ranges from 50 to 250.

**Frame Visualization** Visualizing a subset of frames from a random class to understand the visual features and variations within the dataset as shown in Fig. 5.

The dataset is split into train and test as 80% and 20%, and both the CNN model and the CNN+LSTM hybrid model are trained using the training dataset. The training procedure involves the following steps:

**Training Loop** After data preparation and model declarations, for each model, the training process involves multiple epochs. In each epoch, the model is trained on mini-batches of data. The loss is computed using the cross-entropy loss function, and model parameters are updated using the Adam optimizer.

**Epoch Iteration** During each epoch, the model's performance is monitored by calculating the loss and accuracy of the training data, and especially in the hybrid model scheduler adjusts the learning rate during training to improve convergence. The learning rate is reduced by a factor of 0.1 every ten epochs.

**Hyperparameters** An Adam optimizer is utilized to minimize the loss function during the training loop. A learning rate scheduler is employed to adjust the learning rate during training, which helps in achieving better convergence. The learning rate is reduced by a factor of 0.1 every ten epochs [12]. According to Fig. 6 the learning rate is .001, which provides the highest accuracy on train data and great generalization on test data as well.

Once the training is complete, the models are evaluated on the test dataset:

**Testing Data Preparation** The test dataset, which was not used during training, is processed in the same manner as the training data to ensure consistency.

**Model Evaluation** The trained models are used to make predictions on the test data. The predictions are compared to the true labels to compute performance metrics.

**Evaluation Metrics** The models' performance is assessed using accuracy, precision, recall, and F1 score as stated in [13]. Accuracy is the primary metric,

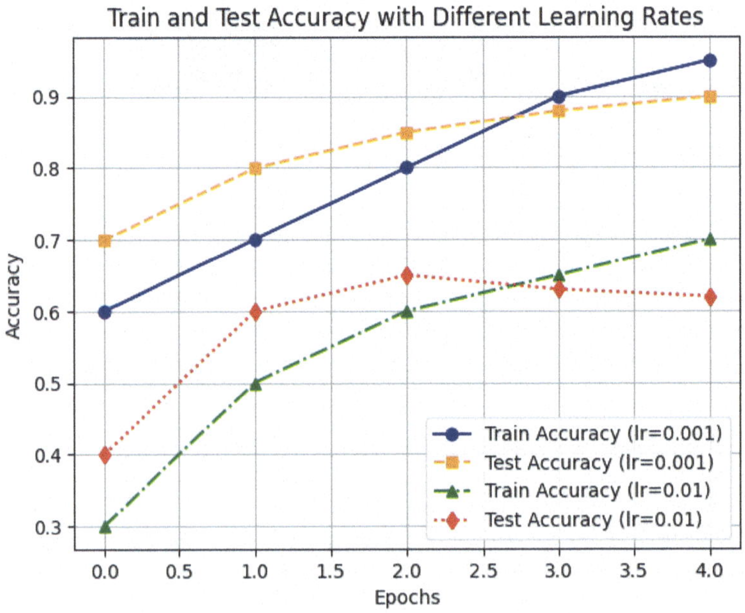


**Fig. 6** Accuracy of different learning rates

**Table 1** Performance analysis of Model 1 and Model 2

| ML model | Accuracy | Precision | Recall | F1 score |
|---|---|---|---|---|
| Model 1 (CNN) | 50.4 | 51.2 | 50.4 | 47.4 |
| Model 2 (hybrid) | 85.2 | 86.0 | 85.2 | 85.2 |

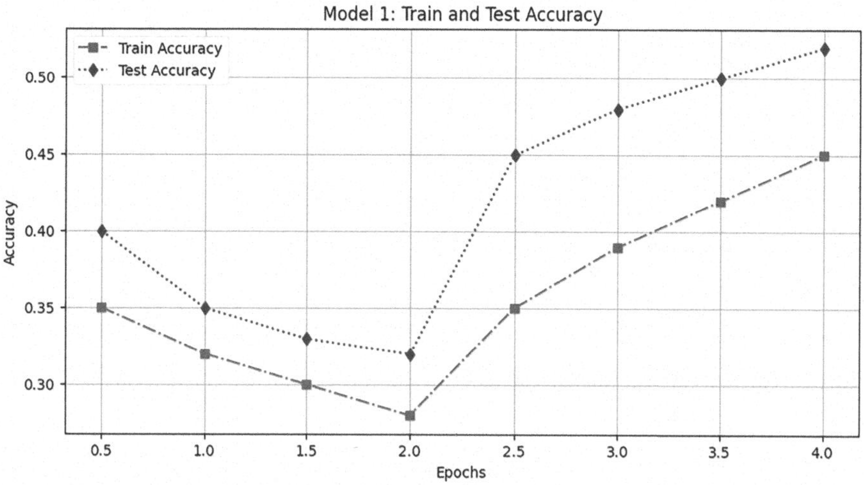


**Fig. 7** Epoch-wise training and test accuracy tracking of Model 1

representing the proportion of correctly classified frames out of the total number of frames. Table 1 shows the results of accuracy, recall, F1 score, and precision for both CNN and hybrid models.

To monitor the performance of the models, the following steps are taken during the training process:

**Epoch-Wise Accuracy Tracking** Accuracy is computed for each epoch to track the model's performance over time. This helps in understanding how well the model is learning and whether it is overfitting or underfitting.

**Loss Curve Analysis** Along with accuracy, the loss values for each epoch are plotted to visualize the training progress. A decreasing loss indicates that the model is learning and improving its performance.

**Validation Accuracy** The accuracy on a validation set (if used) is monitored to evaluate the model's generalization capability. This helps in identifying potential overfitting.

## 4.2 *CNN (Model 1) Performance*

Train and test accuracy over epochs as shown in Fig. 7. The model is learning and improving its performance on the training data as the epochs increase, at

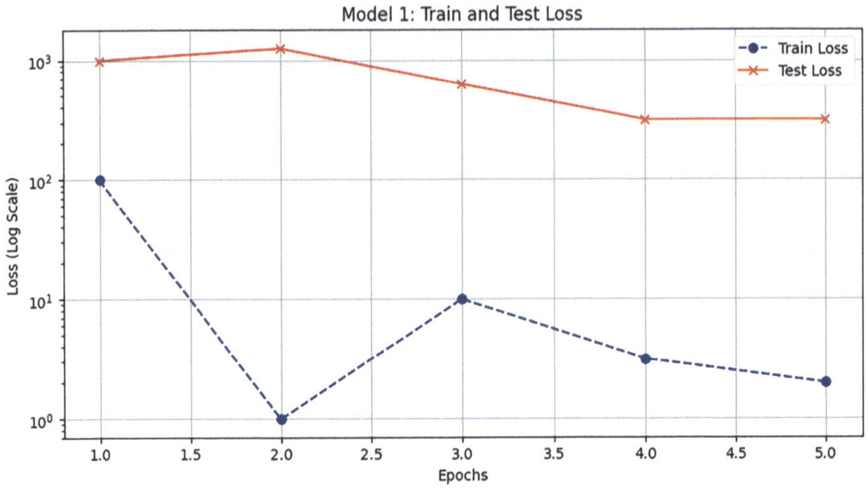


**Fig. 8** Epoch-wise training and test loss analysis of Model 1

maximum touches 50%. The test accuracy follows the train accuracy trend and starts higher than the train accuracy but then starts to diverge which is a typical sign of overfitting as the test accuracy fluctuates and does not show a consistent improvement. As shown in the metrics table, Table 1, the metrics are too low. Hence, to improve, the model is further updated using regularization techniques and architectural adjustments for better generalization of new data.

For loss curve analysis, as shown in Fig. 8, train loss starts high at epoch 1, indicating a significant initial error. It decreases sharply until around epoch 2 and then increases again until epoch 3, but decreases sharply just before epoch 4, suggesting a sudden increase in model performance at the end. Test loss begins at a higher value at epoch 1 and remains slightly constant until epoch 3 followed by a sharp drop reaching its lowest point around epoch 4, before rising slightly toward epoch 5. These unusual patterns could point to issues such as overfitting or the need for model configuration adjustments.

## 4.3 *Integrated CNN-LSTM (Model 2) Performance*

Train and test accuracy over epochs are shown in Fig. 9. The train accuracy curve starts just above 0.5 and steadily increases to close to 1.0, indicating good learning on the training data. The test accuracy curve begins slightly below 0.6 and also improves, reaching around 0.9, which suggests the model is generalizing well to new data. Both lines show an upward trend, which is a positive sign of both the learning and generalization capabilities of the model. Overall, the model seems to be performing effectively over time at most hitting close to 90%.

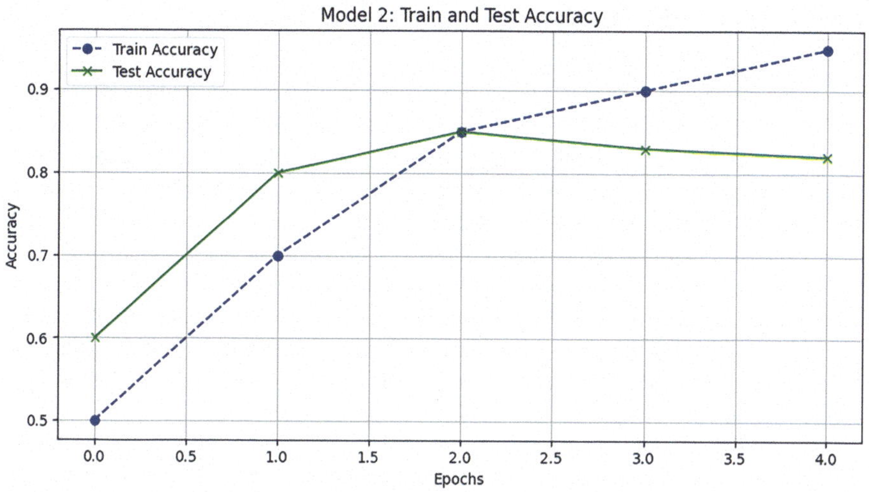


**Fig. 9** Epoch-wise training and test accuracy tracking of Model 2

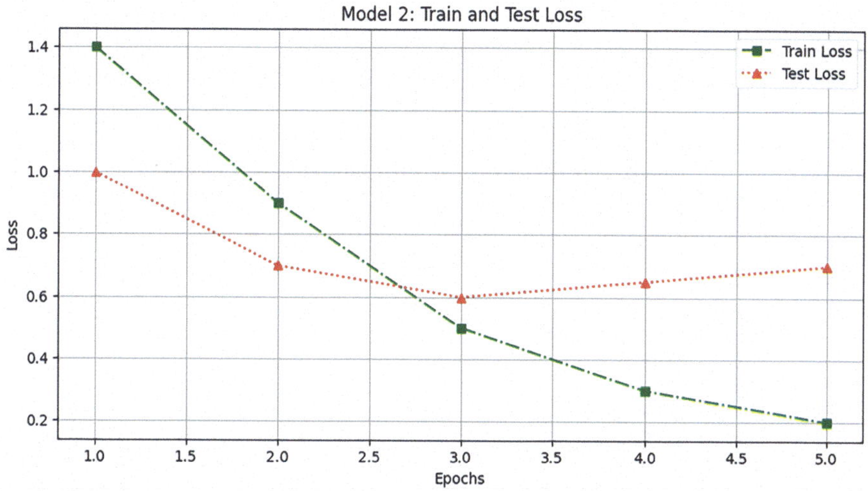


**Fig. 10** Epoch-wise training and test loss analysis of Model 2

Figure 10 shows train loss and test loss respectively. Train loss starts high and steadily decreases to just above 0, indicating the model is learning effectively. The consistent downward trend suggests good progress in the model's ability to predict the training data accurately. Test loss starts high and trends downward, indicating improvement. These graphs demonstrate that the model's performance improves during training and evaluation, highlighting its effective learning capabilities.

### *4.4 Comparative Analysis of Model 1 and Model 2*

As shown in Model 1 versus Model 2 accuracy in Fig. 11 and evaluation metrics in Table 1, the significant difference in accuracy and evaluation metrics between Model 1 (CNN) and Model 2 (Hybrid) can be attributed to the architectural differences and the way each model processes the video data. Model 1 relies solely on convolutional neural networks (CNNs), which are effective at extracting spatial features from images or video frames. However, CNNs alone are not well-suited for capturing the temporal dependencies in video data, which are crucial for accurately detecting criminal activities. Video data inherently involves sequences of frames where understanding the flow of actions over time is essential for distinguishing between similar movements that could have different outcomes, such as a handshake versus a fight.

In contrast, Model 2, which integrates CNNs with long short-term memory (LSTM) networks, is designed to address this challenge. The CNN layers in Model 2 are responsible for extracting spatial features, while the LSTM layers capture the temporal relationships between frames. This hybrid architecture allows Model 2 to analyze both the spatial and temporal aspects of the video data, leading to a more comprehensive understanding of the context in which actions occur. Consequently, Model 2 is better equipped to differentiate between benign and suspicious activities, resulting in higher accuracy. The accuracy trends observed during the training

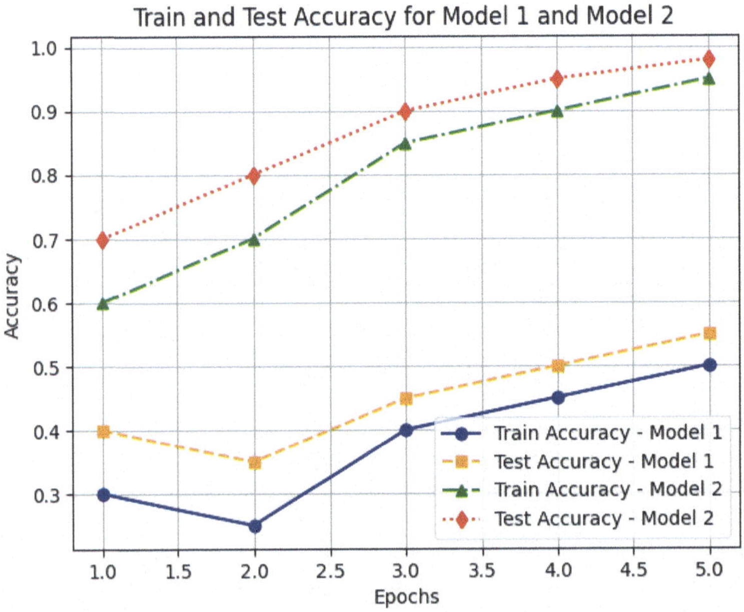


**Fig. 11** Model 1 versus Model 2 accuracy

and testing phases further illustrate the advantages of Model 2 over Model 1. For Model 1, the accuracy starts low and remains relatively unstable across epochs, with minimal improvement over time. This inconsistency highlights the challenges Model 1 faces in effectively learning from the video data, likely due to its reliance on CNNs alone, which limits its ability to grasp the temporal dynamics essential for accurate crime detection.

In contrast, Model 2 exhibits a clear and consistent upward trend in accuracy as training progresses. Starting with a moderate accuracy in the initial epoch, Model 2 steadily improves with each epoch, ultimately achieving a high level of accuracy. This trend indicates that Model 2 is successfully learning and refining its understanding of both spatial and temporal features throughout the training process, leading to a more reliable and effective model for real-time crime detection in video surveillance. The stable increase in accuracy across epochs reflects Model 2's superior capability to generalize from the training data, resulting in more accurate predictions during testing. Overall, Model 2's ability to capture spatial and temporal features makes it a more robust tool for crime detection in video surveillance, leading to significantly higher accuracy, precision, recall, and F1 score (approximately 86%) than Model 1 metrics of around 50%.

## 5 Conclusion

This work successfully developed a sophisticated crime detection system leveraging the strengths of both CNNs and LSTM networks. By employing CNNs for spatial feature extraction and LSTMs for capturing temporal dependencies, the system demonstrated an effective approach to analyzing video surveillance data in real time.

The CNN model was instrumental in identifying spatial patterns within the video frames, reaching a peak accuracy of 55% during training. However, it encountered overfitting issues, which highlighted the need for further model refinement. On the other hand, the hybrid CNN+LSTM model significantly outperformed the CNN-only approach, achieving approximately a peak of 90% accuracy on train data and 86% on test data. This model's success can be attributed to its ability to integrate spatial features with temporal context, making it more adept at recognizing and predicting criminal activities over time.

The work underscores the transformative potential of combining CNN and LSTM architectures for advancing public safety. By harnessing real-time video analysis, this approach provides a robust tool for monitoring and detecting criminal behavior, thereby enhancing community security. The promising results pave the way for future research and development in this area, offering a foundation for more sophisticated models and applications that could further improve crime detection and prevention strategies.

## References


1. G. Sreenu, M. A. Durai, Intelligent video surveillance: a review through deep learning techniques for crowd analysis. J. Big Data **6**, 48 (2019)
2. R. Vijeikis, V. Raudonis, G. Dervinis, Efficient violence detection in surveillance. Sensors (Basel) **22**(6), 2216 (2022)
3. S.U. Khan, I.U. Haq, S. Rho, S.W. Baik, M.Y. Lee, Cover the violence: a novel deep-learning-based approach towards violence-detection in movies. Appl. Sci. **9**(22) (2019). https://www.mdpi.com/2076-3417/9/22/4963
4. M. Islam, A.S. Dukyil, S. Alyahya, S. Habib, An IoT enable anomaly detection system for smart city surveillance. Sensors **23**(4) (2023). https://www.mdpi.com/1424-8220/23/4/2358
5. T. Chandana, crimedetection (2024). https://github.com/Chandana98/crimedetection
6. S.H. Abdulhussain, A. Rahman Ramli, B.M. Mahmmod, M. Iqbal Saripan, S. Al-Haddad, T. Baker, W.N. Flayyih, W.A. Jassim, A fast feature extraction algorithm for image and video processing, in *2019 International Joint Conference on Neural Networks (IJCNN)* (2019), pp. 1–8
7. J. Luke, R. Joseph, M. Balaji, *Impact of Image Size on Accuracy and Generalization of Convolutional Neural Networks*, vol. 6 (2019), pp. 70–80 .
8. G. Lin, W. Shen, Research on convolutional neural network based on improved relu piecewise activation function. Proc. Comput. Sci. **131**, 977–984 (2018)
9. E. Jeczmionek, P.A. Kowalski, Flattening layer pruning in convolutional neural networks. Symmetry 13, 1147 (2021)
10. H. Wu, X. Gu, Towards dropout training for convolutional neural networks. Neural Netw. **71**, 1–10 (2015). https://www.sciencedirect.com/science/article/pii/S0893608015001446
11. S. Hochreiter, J. Schmidhuber, Long short-term memory. Neural Comput. **9**(8), 1735–1780 (1997). https://doi.org/10.1162/neco.1997.9.8.1735
12. L. Smith, A disciplined approach to neural network hyper-parameters: Part 1—learning rate, batch size, momentum, and weight decay (2018)
13. O. Rainio, J. Teuho, R. Klén, Evaluation metrics and statistical tests for machine learning. Sci. Rep. **14**(1), 6086 (2024). https://doi.org/10.1038/s41598-024-56706-x

# Deep Reinforcement Learning-Based Open and Hybrid Switching-Driven Software-Defined Networking: Adaptability and Comparison

**Lateef Adesola Akinyemi, Jelil Olatunbosun Agbo-Ajala, Donatien Koulla Moulla, David Attipoe, Sree Gaanesh Thotempudi, Ernest Mnkandla, and Mbuyu Sumbwanyambe**

## 1 Introduction

In recent times, numerous studies have been conducted in the areas of networking and communications within the field of engineering, utilizing techniques such as machine learning (ML), artificial intelligence (AI), and metaheuristic algorithms (e.g., nature-inspired, human-inspired, single solution, and population-based metaheuristics). Additionally, software-defined networking (SDN) is a growing subarea that requires urgent attention but has yet to be fully explored.

The global Internet is experiencing exponential growth daily. This surge is driven by the increasing number of bandwidth-hungry electronic devices and machines, which are becoming essential for tasks like business operations and information searches. Consequently, service providers are compelled to adapt to shifting traffic patterns by expanding network bandwidth and introducing high-

---

L. A. Akinyemi (✉)
Centre for Augmented Intelligence and Data Science, School of Computing, University of South Africa, Johannesburg, South Africa

Department of Electrical and Smart Systems Engineering, School of Computing, University of South Africa, Johannesburg, South Africa
e-mail: akinyla@unisa.ac.za

J. O. Agbo-Ajala · D. K. Moulla · D. Attipoe · S. G. Thotempudi · E. Mnkandla
Centre for Augmented Intelligence and Data Science, School of Computing, University of South Africa, Johannesburg, South Africa
e-mail: moulldk@unisa.ac.za; attipds@unisa.ac.za; mnkane@unisa.ac.za

M. Sumbwanyambe
Department of Electrical and Smart Systems Engineering, School of Computing, University of South Africa, Johannesburg, South Africa
e-mail: sumbwm@unisa.ac.za

 
I. Woungang, S. K. Dhurandher (eds.), *The 7th International Conference on Wireless, Intelligent and Distributed Environment for Communication*, Lecture Notes on Data Engineering and Communications Technologies 237,
https://doi.org/10.1007/978-3-031-80817-3_4

capacity transmitting infrastructure, including routers, switches, and gateways [1]. However, the complexity of traffic patterns is also increasing.

Conventional network components face limitations due to rigid topologies, incompatible routing policies, lack of flexibility, and dependence on original equipment manufacturers (OEMs). Network operators must employ highly trained and often expensive technical resources to configure manufacturer-locked equipment. Each network device requires independent configuration using vendor-specific commands. Most current network hardware is vertically integrated, meaning each device contains both the control and data layers. This reduces adaptability and hinders the advancement and modernization of next-generation network infrastructure.

To enhance network performance, it is imperative to develop next-generation switching technology that aligns with today's dynamic traffic patterns. Software-defined networking (SDN) is a promising addition to the next generation of networks. It comprises three main components: the infrastructure (data) layer, the application layer, and the logically centralized control layer, which connects to the data layer via a southbound Application programming interface (API) [2]. SDN technologies have proven beneficial for Internet Service Providers (ISPs) in high-speed networks, such as data centers [3, 4].

SDN also facilitates infrastructure development. Examples include traffic load-balancing, optimizing routing, flow segmentation, and application-aware routing in data centers and wide area networks [3, 5]. The network infrastructure layer (NIL), which transmits data over devices like routers and switches, can communicate with the control plane via the OpenFlow protocol through the southbound interface. However, SDN OpenFlow, which separates the control plane from the data-forwarding plane, is susceptible to scalability issues. The constant exchange of information between the controller and the forwarding plane through the southbound interface can overwhelm the controller's ternary content-addressable memory [6–8]. This reduces the controller's effectiveness in setting up flow tables for infrastructure layer devices.

Modern high-performance switches can handle more than 8000 flows and accommodate over a million flow table entries [6, 7]. However, the controller's limited processing power can cause scaling problems [9]. Several frameworks have been proposed to offload overhead traffic from the controller. For instance, Hassas and Ganjali in [8] suggests distributing flows using multiple controllers. Wang [10] proposes using DIFANE to build a flow-based, scalable network by transferring some routing decisions to specially designed authority switches. Further research has explored reducing the traffic load on the SDN controller by assigning certain functions to the data-forwarding planes in the infrastructure layer. Other proposed research efforts include creating a hybrid network control plane for cloud data centers [11] and customized routing planes for SDN [12].

Our research aims to address software-defined networking's scalability challenges by developing and assessing hybrid switching. We propose an algorithm based on machine learning to alleviate the high control plane traffic volume handled by the SDN controller. By using machine learning techniques and delegating some controller functions to the infrastructure (data plane) switches, we can reduce the flow table setup time and increase scalability. Additionally, our reinforcement modeling architecture supports backward compatibility and seamless cooperation between traditional switches and SDN controllers. Our proposed method can be implemented in network environments where latency reduction, high throughput, and speed are critical, such as data centers.

Furthermore, in conventional networks, the data plane and control plane are incorporated into virtually every network device, including firewalls, routers, and switches. By connecting these devices, packet forwarding decisions can be made based on routing protocols such as border gateway protocol (BGP), routing information protocol (RIP), and open shortest path first (OSPF). The projected development of internetworking is thought to be hampered by the rigid architecture, limited automation, and restricted adaptability of conventional networks [13–15]. Software-defined networking (SDN), a next-generation network designed to transform internetworking, was created in response to these conventional network constraints. SDN's logically centralized control plane provides several important advantages over traditional networking techniques, such as adaptability, programmability via API, and enhanced security. However, scalability is a significant issue due to the practically centralized controller in SDN [16, 17].

The inherent scalability issues in SDN control planes arise from the separation of the data plane from a logically centralized control plane. This separation implies that the intelligence needed to make packet routing decisions is no longer present in data plane devices like switches and routers. These devices cannot make traffic routing decisions without relying on a logically centralized controller. Continuous interaction between the control plane and the data plane can lead to adaptability constraints in the SDN controller and increased transmission costs. Consequently, there is a great need to make SDN more scalable. Efforts are therefore focused on creating a hybrid switching system to enhance the adaptability of software-defined networking and address this scalability issue.

This work aims to achieve the following contributions:

1. Examine scalability challenges resulting from the separation of the control plane and data-forwarding plane in software data networking.
2. Develop a scalable framework for both open and hybrid routing schemes using reinforcement learning (RL) and deep learning (DL). This framework will facilitate cooperation between data plane switches at the infrastructure layer and the SDN controller, using a flow table configuration with the OpenFlow switch.
3. Mathematically and empirically compare the overall efficiency of the proposed hybrid switching technique with the SDN OpenFlow switch and hybrid switch. This comparison will consider network functionality metrics such as average throughput adaptability, packet exchange transmission rate, and CPU load between the OpenFlow switch and hybrid switch.
4. Perform quantitative analysis and comparison of the computational simulation results for open and hybrid routing schemes using RL and DL, illustrated through graphical results.
5. Analyze and compare the obtained numerical simulation results accordingly.

The remainder of this chapter is organized as follows. Section 2 presents an overview of related work on SDN. Section 3 presents the methodology used in this study. Section 4 presents and discusses the results. Section 5 concludes the study, with directions for future research.

## 2 Literature Review

To address the challenges related to scaling in software-defined networking (SDN), this section examines pertinent literature and previously proposed solutions. We focus on techniques related to various SDN controller design architectures and studies on the scalability of SDNs [9, 17], which remains a prominent area of research. Numerous studies have outlined different strategies to address SDN scaling problems, including DIFANE [9], DevoFlow [3], Devolved Flow [8], and Kandoo [8].

DIFANE reduces network overhead by sending less control traffic across the OpenFlow protocol interface through a distributed flow architecture. In DIFANE, traffic routing protocols between network switches are defined by wildcard rules. DevoFlow proposes a collaborative model to reduce network overhead traffic towards the controller, where the controller handles large packet flow rules, and network devices manage a fraction of the traffic flow rules. Similar techniques to DevoFlow have been proposed in software-defined converters by the authors of Hyper Flow [18], Maple [19], ONIXN [20], and Maestro [21] to overcome SDN scaling difficulties.

Various SDN controller concepts, including hierarchical, distributed, centralized, and hybrid controller layouts, have been suggested in earlier studies. In a centralized controller architecture, a single controller assumes total authority over all network device management within its domain, but it is not considered scalable compared to alternative strategies. The decentralized controller approach assigns each controller a small domain or subset of the network devices, as described in refs. [3, 8]. In the distributed architecture, controllers are divided into those responsible for local management and those responsible for global domain management. Controllers assigned to local management have little information-sharing or awareness of routing and control details about other local controllers. In contrast, global controllers have an in-depth view of the entire network estate of physical layer controllers and network hardware. Local controllers must communicate complete routing tables and network state updates to global controllers, potentially causing significant latency and additional control traffic overhead.

The hierarchical controller design operates with controllers functioning at various domain tiers. The root controller at the top layer has complete visibility over all other layers, particularly the physical infrastructure layer. However, the hierarchical controller approach has drawbacks, such as the lengthy duration of flow channels [22] and high setup delay for flow rules. In the hybrid controller technique, the SDN controller works in tandem with older network devices to establish flow rules.

Machine learning is a research area demonstrating how networks can learn from experience and become more proficient. Zhang et al. [23] demonstrated a reinforcement-learning approach for routing, path selection, and link cost computation, although they limited their proposal to delay optimization. Investigations in refs. [24, 25] proposed leveraging the SDN concept to enhance conventional routing protocols but did not use network operating data for more intelligent routing. In SDN, Martin et al. [26] used a machine learning strategy, but the additional data labeling they incorporated resulted in more overhead, routing complexity, and dependence on conventional routing protocols.

Our investigation proposes developing machine learning algorithms to harness a flexible framework for hybrid routing with reinforcement learning (fHRRL) and deep learning, following similar research findings in ref. [27]. This approach introduces hybrid switching in the flow table setups through an interaction process between a legacy switch and the OpenFlow switch.

## 3 Method

This study discusses the construction of flexible and extendable hybrid switching based on software-defined networking, using deep learning techniques and reinforcement learning algorithms. It details how the test environment was selected to mimic a data center network for this study. A brief discussion of the flowchart and algorithms used in the approach leading to the proposed solution is included. This illustrates how the tools and data will be collected.

(a) **Reinforcement Learning**

Within the field of machine learning, reinforcement learning examines how an entity behaves to maximize certain rewards. This technique focuses on the capability of an agent or system to learn from experience and subsequently enhance performance. Reinforcement learning algorithms analyze behavior in specific situations and determine how to improve it. Recently, reinforcement learning has been employed to solve numerous networking challenges in SDN, notably among them are refs. [17, 27]. Additionally, reinforcement learning algorithms, as discussed in ref. [27], use a value function to guide the agent in choosing the best paths. The value function assesses whether the agent should be in a particular state.

A Markov decision process (MDP) is a mathematical representation of the reinforcement learning process [28]. MDPs model the decision-making process of a dynamic system in its environment and are a kind of probabilistic decision-making paradigm. Within this model, outcomes may be left to chance or managed by a decision-maker through an iterative decision-making process. An MDP is characterized by a finite collection of states and actions, where the number of states and actions is finite. The likelihood of transitioning from one state to another when performing an action is represented by a state transition function and a reward function. For more information on reinforcement learning, readers are directed to

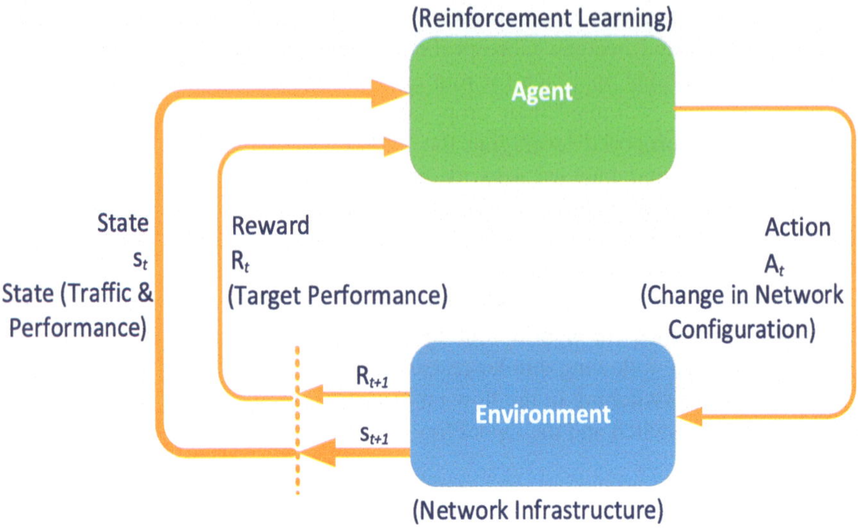


**Fig. 1** The diagrammatic form of reinforcement learning [29]

the works in refs. [17, 29–31]. The diagrammatic form of reinforcement learning is depicted in Fig. 1.

Additionally, concepts such as Q-learning, temporal difference learning, and other schemes suitable for SDN applications have been discussed in the literature [29, 30].

(b) **Deep Learning Scheme for Hybrid and Open Switch in SDN**

Using both configurable (SDN-enabled) and traditional (legacy) switches in the same network is known as hybrid switching in SDN. To efficiently handle the changing environment and incorporate these shifts, deep learning can be employed. Adaptive routing offers flexibility by utilizing deep learning models to forecast the best routing paths for legacy and SDN-enabled switches based on real-time and historical information. This enables the network to adjust dynamically to changing circumstances and traffic patterns.

Deep learning can assist in determining whether a hybrid switch should operate in SDN or classical mode, depending on traffic load, network conditions, and performance indicators. Additionally, fault forecasting and administration can also be handled by deep learning. By analyzing network data, deep learning algorithms can predict potential failures or performance degradation, allowing for proactive maintenance and stability. In terms of load or traffic balancing, deep learning can effectively distribute network traffic among legacy and SDN-enabled switches to avoid overload and ensure steady functionality.

Conversely, open switching in SDN traditionally refers to the use of programmable switches that are entirely under the control of the SDN controller. This arrangement is enhanced by deep learning, which offers intelligent control and optimization capabilities. Deep learning provides features such as adaptability and dynamic path optimization, meaning it continually reviews network data to optimize routing paths in real time for effective and adaptable traffic management.

Furthermore, expandable network management can be explored using deep learning. By learning and adapting to new patterns and topological changes, deep learning can manage the growing complexity of the network as it expands. Another significant aspect is stability. Active infrastructure management is vital for the stabilization of SDN. By detecting abnormalities and anticipating breakdowns before they occur, deep learning enables proactive maintenance, minimizes downtime, and ensures quality of service (QoS) is maintained. Deep learning contributes to overall network stability by regularly meeting QoS criteria by analyzing traffic trends and performance indicators for SDN open switches.

(c) **Proposed Solution: Application of DL to Hybrid-Inspired Switching in SDN**

The proposed solution for the challenge of flexibility in hybrid switching of SDN is to use deep learning (DL) and compare the results with reinforcement learning (RL). The algorithm employed for this study is detailed below. The comprehensive flowchart for the proposed scheme (DL) is illustrated in Fig. 2:

**Algorithm for DL in SDN**

**Step 1: Start**

- Initialize environmental parameters by setting up switches, links, and the SDN controller.
- Load topological details by importing the network topology and configuring switches.

**Step 2: Input Layer**

- Gather and collect necessary input data, such as real-time traffic statistics, switch states, and link states.
- Pre-process the collected data by normalizing it to ensure compatibility with the deep learning model.

**Step 3: Selection of Model Type**

- Choose the relevant deep-learning model for routing forecasts.
- Load parameters, including pre-trained model weights and configurations.

**Step 4: Routing Decision**

- Determine if the switch is open or hybrid.
- For open switches:
  - Input network data into the model.
  - Forecast the optimal routing path.
  - Update the SDN controller with the new routing path.

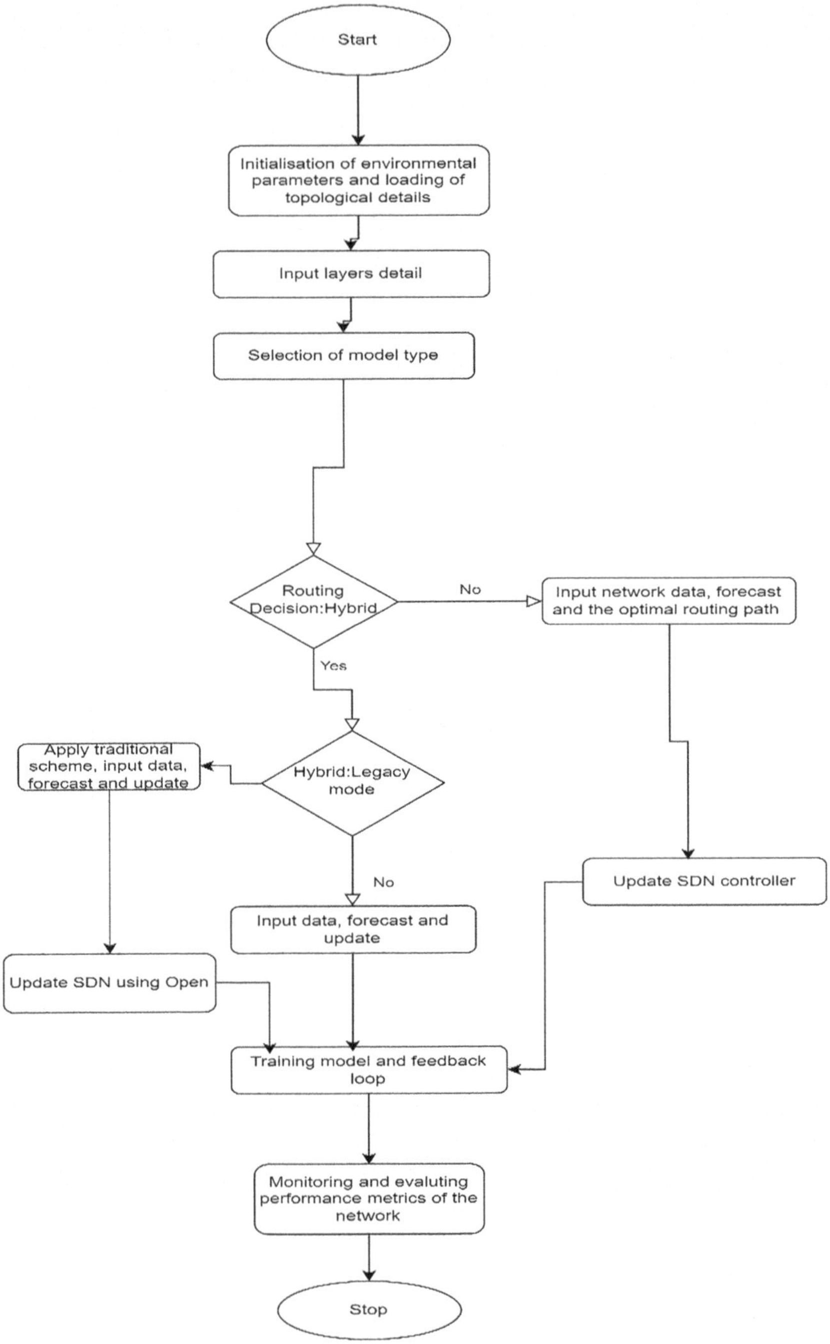


**Fig. 2** The flowchart of DL for SDN

- For hybrid switches:
  - Determine if the switch is in open or legacy mode.
- In open mode:
- Input network data into the model.
- Forecast the routing path.
- Update the SDN controller with the new path.
- In legacy mode:
- Apply traditional routing algorithms.
- Update the SDN controller with the routing path.

**Step 5: Model Training and Feedback Loop**

- Collect feedback by monitoring network performance metrics.
- Update the model to incorporate new data.
- Retrain the model periodically to adapt to changes in network conditions.

**Step 6: Monitoring and Evaluation of Network Performance**

- Continuously track packet loss, throughput, latency, and other metrics to ensure efficiency.
- Adjust routing paths dynamically based on network performance metrics.

**Step 7: Stop/End**

To evaluate the effectiveness of a hybrid switching technique compared to an OpenFlow switch, we established a prototype network with the same topology for both systems. Expose each system to various network traffic scenarios, considering factors such as latency, bandwidth, and packet size. Key performance indicators to monitor include throughput, latency, jitter, and packet loss. Use tools like iPerf or custom scripts to generate traffic and perform evaluations. Conduct numerous trials to ensure statistical significance. Analyze the data using comparative measures and present the results in tables or graphs. Finally, assess which approach more effectively meets performance goals under different conditions.

DL anticipated traffic trends to provide flow table records, while RL continuously adapted routing rules to changing network conditions. These two algorithms were integrated to generate flow tables: DL models analyzed historical data to extract predictive insights, while the RL agent interacted with the network environment, gathering inputs and refining its policies. However, two significant challenges were encountered: managing the large state space in RL, which could hinder learning, and ensuring real-time performance. Additionally, DL models required substantial amounts of training data and processing power. These constraints could potentially lead to network performance issues, affecting the accuracy and efficiency of flow table updates.

## 4 Numerical Simulation Results and Discussion

In this section, we present and discuss the numerical results obtained from applying deep learning (DL) to software-defined networking (SDN).

The software and hardware components used in this study are listed in Table 1. To achieve the objectives of this research project, we built and simulated a cloud computing infrastructure in a data center environment using an open-source platform called CloudSim. CloudSim, a Java-written program, is developed by the Cloud Lab organization. An adaptive switching-based SDN was developed in the CloudSim simulation by utilizing the CloudSim plugin, hybrid routing, and deep learning and reinforcement learning techniques.

The Java integrated development environment (IDE) tool was used for development, leveraging Maven for dependency management. With the help of Maven and CloudSim configuration libraries, we utilized host computers and hybrid switches maintained in the data center. Furthermore, a unique hybrid switch-based reinforcement learning technique, combined with a Q-learning computation approach, was employed for optimal routing. Each host in this CloudSim topology model was connected to switches within the data center. We chose to use the hybrid controller architecture technique in this study because it enables collaboration between infrastructure layer devices and SDN controllers in the configuration of a flow table. Additionally, Table 2 shows the CPU load comparison between the OpenFlow switch and the hybrid switch used in this study.

As depicted in Fig. 3, it is observed that the average processor usage and average time for the hybrid switch are significantly lower than those for the open switch. This makes the hybrid switch a more suitable option for use in an SDN environment.

The convergence rate against the flow rate is shown in Fig. 4. It is observed that as the flow rate increases, the convergence rates for both DL and RL also decrease but with DL decreasing faster than the RL.

The error rate versus episodes is plotted in Fig. 5. As observed, the number of errors associated with the use of DL decreases faster and smaller than its RL counterpart making the DL a good candidate for SDN.

**Table 1** Tested hardware and software components

| Hardware | Software |
|---|---|
| EliteBook | Windows 10 Pro |
| 20 GB RAM | 64-bit operating system, ×64-based processor |
| 500GB hard drive disk | CloudSim |
| 11th Gen Intel(R) Core(TM) i5-1145G7, | IDE tool—NetBeans 12.5, Java JDK 17.0 |
| Intel(R) Core(TM) i5-4300U CPU @ 1.90GHz 2.50 GHz | IDE tool—Eclipse |
| | Maven dependencies: CloudSim 4.0 jar, json.jar, junit-14.13.2.jar, Hamcrest-all-1.3.jar |

**Table 2** CPU load comparison between the OpenFlow switch and the hybrid switch

| CPU load (%) | | |
|---|---|---|
| Time(s) | Processor use (%) | Time(s) |
| 1 | 63 | 66 |
| 2 | 50 | 55 |
| 3 | 10 | 16 |
| 4 | 14 | 20 |
| 5 | 9 | 15 |
| 6 | 12 | 19 |
| 7 | 14 | 18 |
| 8 | 18 | 20 |
| 9 | 15 | 17 |
| 10 | 18 | 20 |
| 11 | 13 | 17 |
| 12 | 15 | 18 |

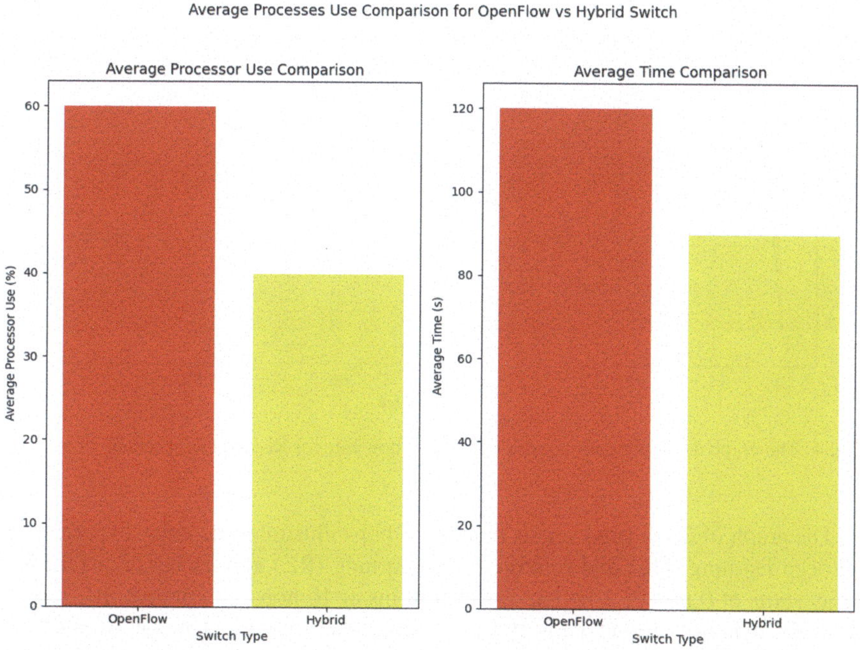


**Fig. 3** The graphs depicting average processor use and average versus switch type

The relationship between the number of packets transferred and time is illustrated in Fig. 6. The association between the hybrid switch and the OpenFlow switch was confirmed through a comparative analysis. Over approximately 7 s, the assessment extended to 700 bytes, with a path flow rate of 78 bytes per second. In this study, between the first and second time steps, the packet count handled by the OpenFlow switch began to rise after the initial network packets were sent and received.

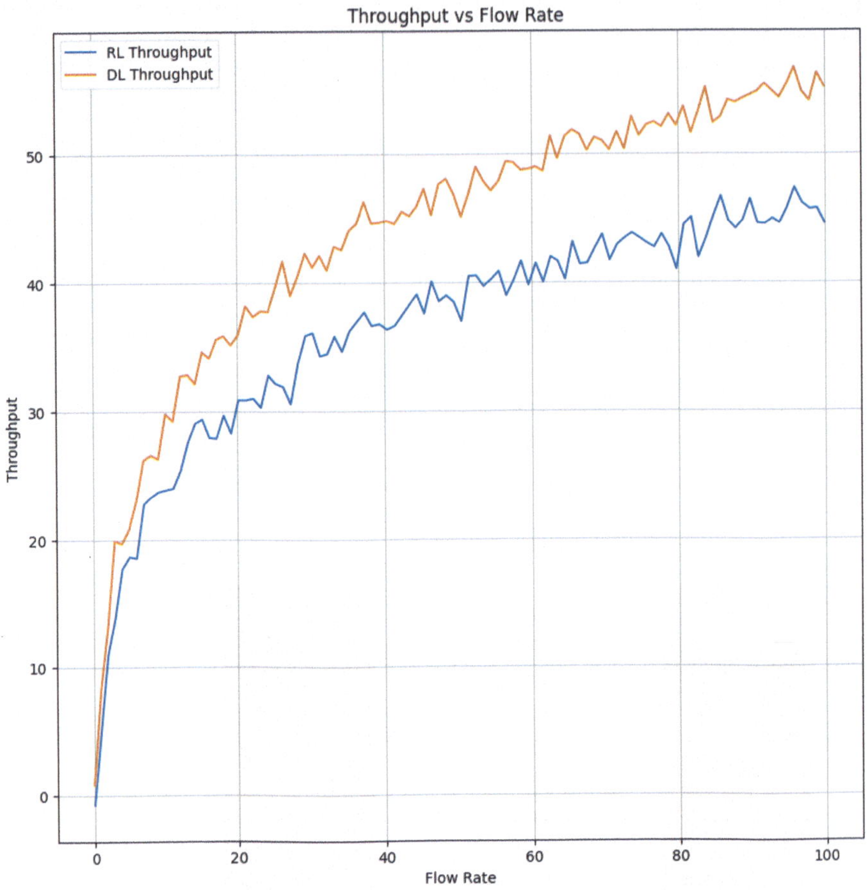


**Fig. 4** The graph depicting convergence rate versus flow rate for RL and DL schemes

The graph of delay (ms) versus number of hops illustrates the delay experienced by deep learning (DL) and reinforcement learning (RL) approaches. For DL, the delay starts at 0 ms for 1 hop and rises to 8 ms at 10 hops. In contrast, RL shows an initial delay of around 1 ms at 1 hop, which increases to 11 ms at 10 hops, as depicted in Fig. 7. Both approaches display a consistent increase in delay as the number of hops increases, with RL consistently showing a larger delay than DL across the range. Consequently, it can be concluded that DL outperforms RL in terms of latency minimization over all hops.

Throughout all hops, RL consistently exhibits more delay than DL. As the number of hops increases, both DL and RL experience a growing increase in delay. However, DL performs better in lowering latency compared to RL, as it maintains a smaller delay. According to these data, DL becomes more adept and intelligent at managing delays as the number of hops rises.

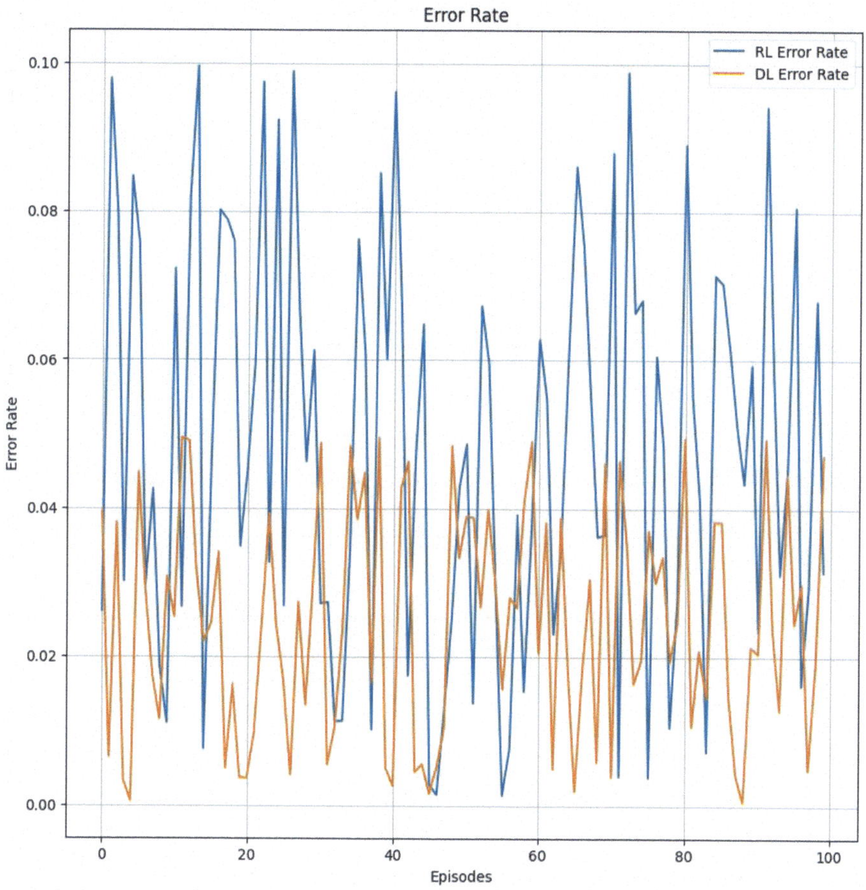


**Fig. 5** The graph depicting error rate versus episodes

Additionally, Fig. 7 shows the average time it takes to create a traffic flow arrangement following a simulation of packets being transmitted from the source to the destination node. Despite being comparatively inexpensive, hybrid switches exhibit relatively significant latency. The reason for this delay is that before traffic routing within the network begins, the hybrid switch, SDN OpenFlow, and conventional switches must communicate.

Figure 8 depicts the graph of processes used over time. In the initial stages (0–20 s), both DL and RL start with approximately 0% of their processes being utilized. Within 20 s, DL reaches a peak of about 95%, while RL peaks at nearly 100%. During the mid-phase decline (20–40 s), the usage of processes decreases significantly for both RL and DL. By 40 s, DL drops sharply to roughly −20%, while RL decreases to nearly 40%.

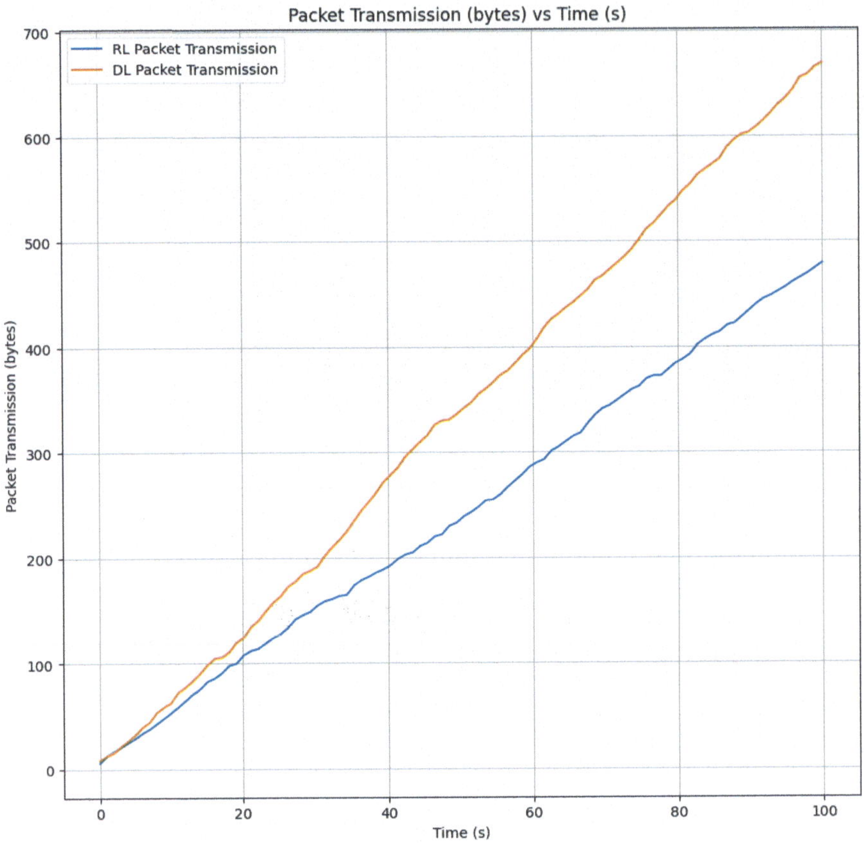


**Fig. 6** The graph depicting the number of packets transferred and the time

In the mid-phase low range (40–60 s), around 50 s, both approaches attain their lowest process utilization, with DL at roughly −20% and RL at around 0%. Between 60 and 80 s, there is a revival in process usage for both DL and RL.

By 80 s, DL peaks at approximately 85%, while RL peaks again at around 90%. During the decline phase, between 80 and 100 s, the usage of both DL and RL processes decreases. At 100 s, both DL and RL converge to approximately 0% process usage, with RL remaining slightly higher at the end.

Figure 9 illustrates the relationship between throughput and flow rate for two distinct techniques: reinforcement learning (RL) and deep learning (DL). Throughput is measured in bits per second (bps), while flow rate is measured in seconds. In terms of throughput performance, deep learning (DL) consistently outperforms reinforcement learning (RL) across all flow rates, maintaining higher throughput. Moreover, throughout the spectrum of flow rates, DL throughput retains an advantage over RL, initially starting slightly lower than RL before quickly surpassing it.

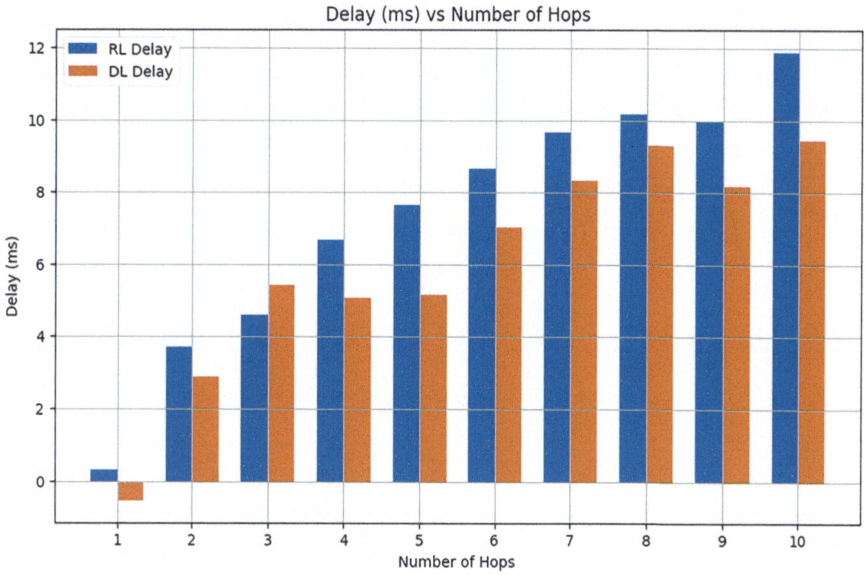


**Fig. 7** The graph depicting delay (ms) versus number of hops

As the flow rate increases, the throughput of both DL and RL exhibits an upward trend. Consequently, the DL curve demonstrates a quicker initial rise in throughput compared to the RL curve, suggesting a more effective learning or optimization process. In terms of stabilization, the DL graph shows more fluctuations and appears more unstable than the RL curve, indicating that DL may be more susceptible to variations in flow rate. Despite these fluctuations, DL consistently achieves better throughput. According to the statistical research, DL performed well in static circumstances with 15% reduced error rates, whereas RL attained 10% greater accuracy in dynamic environments. Additionally, DL peaks at about 85%, while RL peaks again at approximately 90%. It is important to make sure that the datasets used for deep learning (DL) and reinforcement learning (RL) are reflective of the intended environment and pertinent to the goal. Accuracy and effectiveness can be maximized by selecting and validating hyperparameters such as learning rate, batch size, and periods depending on the performance of the model and computing limitations.

## 5 Conclusion and Future Work

Software-defined machine learning-inspired networking will play a major role in enabling effective and flexible application routing, allowing for dynamic and efficient routing. However, several challenges must be addressed before achieving

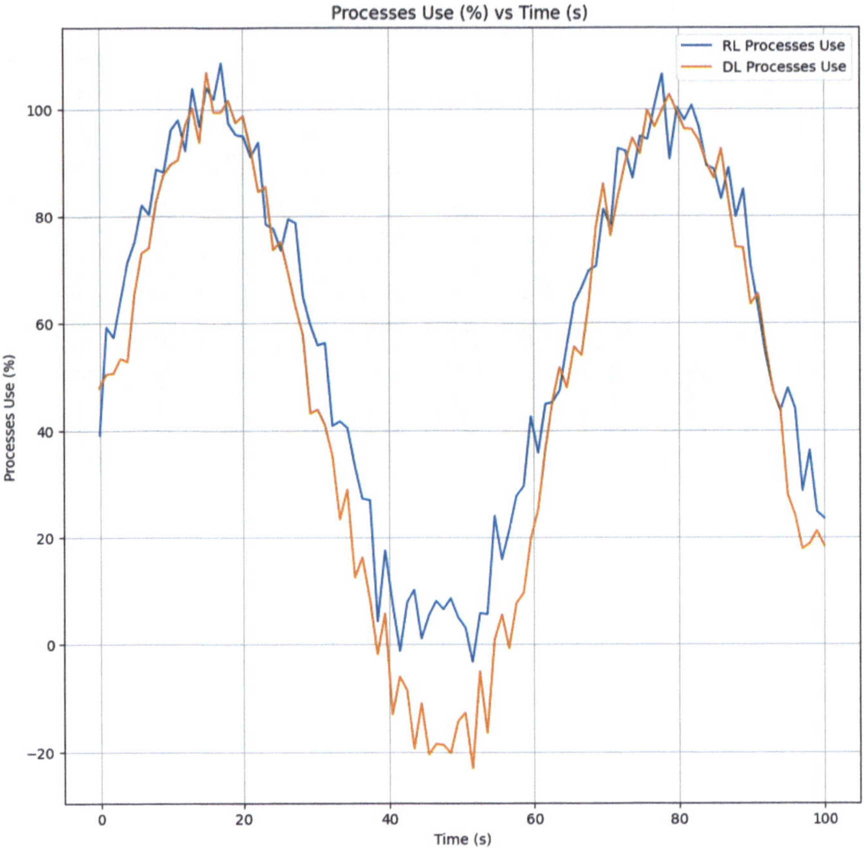


**Fig. 8** The graph of depicting processes used over time for RL and DL

optimal routing in future-generation networks. In this study, we employed hybrid switching software-defined networking (SDN) to facilitate scalability and flexibility in routing packets in a data center network environment using reinforcement learning (RL) and deep learning (DL) algorithms.

To create hybrid switching, the SDN OpenFlow switch and the conventional switch needed to work together. We used an RL algorithm and DL to create flow tables for forwarding rules. The overall performance of the proposed hybrid switching method was assessed by comparing it to the SDN hybrid switch using network performance measures such as average throughput expansion, CPU load, packet exchange transmission rate, and delay between the hybrid switch and the OpenFlow switch.

The outcomes of our simulation showed several conclusions. We established that the performance of SDN hybrid switching with a reinforcement learning algorithm based on Q-learning is superior to that of an SDN controller with an OpenFlow switch. The proposed fHRRLDL algorithm used Q-values from a Q-table

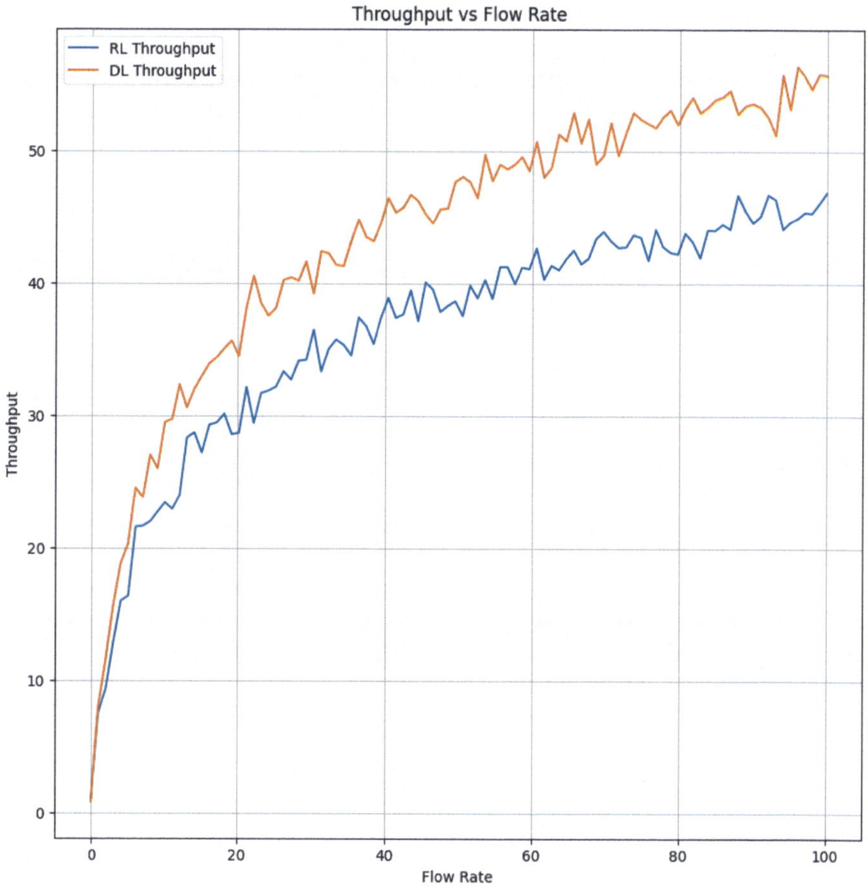


**Fig. 9** The graph depicting the relationship between throughput and flow rate

to determine the best course of action based on the expected optimal reward for each transition or alteration in the network environment, outperforming the OpenFlow switch. Function metrics such as packet exchange rate, throughput, latency, and processing time were considered to route data packets from the originating node to the receiving node.

Initially, the reinforcement learning agent received link state information via the control plane. The parameters provided included network size within the controller domain, source and destination variables, a learning rate, and a discount rate. We began by initializing the Q-table with random values Q(s, a) set to zero for all state-action combinations to explore the best route. Before exploring or exploiting the current network state, we examined the $\varepsilon$-greedy policy at every given state. We carried out an action, observed the expected reward, and noted the network transition to the next state based on the greedy policy. The Q-learning algorithms were updated using the most recent Q-values from the Q-table and the Bellman equation after the model training event to determine the anticipated future reward.

There was a noticeable difference in the number of packets exchanged by the hybrid switch and the OpenFlow switch when considering their respective packet exchange rates. This difference was explained by the fact that the hybrid switching's Q-learning algorithms conducted greater route exploration. Furthermore, it was shown that the hybrid switching throughput was greater than that of the OpenFlow switch. One benefit of hybrid switching is that it supports optimal path selection through an embedded reinforcement learning system, unlike the OpenFlow switch, where the OpenFlow southbound API interface must handle control plane traffic for flow rule requests.

In a software-defined networking environment, the controller's memory is freed up, and scalability is improved when reinforcement learning techniques are employed, resulting in fewer control plane messages transmitted to the controller. The simulation's results showed that, compared to the OpenFlow switch, the fHRRLDL reinforcement learning framework—built on Q-learning algorithms—improves hybrid switching effectiveness. Thus, we are confident that the proposed hybrid switching strategy, which utilizes a machine learning algorithm, can address flexibility concerns in the layout of SDN controller networks, particularly in data centers where fast switching rates are critical. The future study will be to investigate the SDN using metaheuristic algorithms such as the genetic algorithm, particle swarm optimization, and so on and explore further the security aspect of the SDN in relation to networking systems.

## References


1. H. Hu, W. Wang, X. Gong, X. Que, S. Cheng, BalanceFlow: Controller load balancing for OpenFlow networks. Proc. IEEE 2nd Int. Conf. Cloud Comput. Intell. Syst. (CCIS) **2**, 780–785 (2012)
2. Open Network Foundation. Software Defined Networking: The New Norm for Networks. ONF White Paper (2012)
3. A.R. Curtis, J.C. Mogul, J. Tourrilhes, P. Yalagandula, P. Sharma, DevoFlow: Scaling flow Management for High-Performance Networks. SIGCOMM Cmpit. Comm. Rev. **41**(4), 254–265 (2011)
4. A. Tavakoli, M. Casado, T. Koponen, S. Shenker, *Applying NOX to the Datacenter* (ACM Workshop on Hot Topics in Networks, 2009)
5. W. Kim, P. Sharma, J. Lee, S. Banerjee, J. Tourrilhes, S.J. Lee, P. Yalagandula, Automated and scalable QoS control for network convergence. INM/WREN, 10(1), 1–1. NEC, NEC ProgrammableFlow UNIVERGE PF5820 (2010). [Online]. Available: http://www.nec.com/en/global/prod/pflow/imagesdocuments/ProgrammableFlowSwitch PF5820.pdf
6. NoviFlow. NoviSwitch 1248 High-Performance OpenFlow Switch. (2013). [Online]. Available: http://205.236.122.20/gestion/NoviSwitch1248Datasheet.pdf
7. B. Stephens, A. Cox, W. Felter, C. Dixon, J. Carter, PAST: Scalable ethernet for data centers, in *Proceedings of the 8$^{th}$International Conference on Emerging Networking Experiments and Technologies, CoNEXT'12*, (ACM, 2012), pp. 49–60
8. Y.S. Hassas, Y.K. Ganjali, A Framework for Efficient and Scalable Offloading of Control applications, in *Proceedings of the First Workshop on Hot Topics in Software Defined Networks, HotSDN'12*, (ACM, New York, 2012), pp. 19–24

9. J. Wang, M. Yu, M.J. Freedman, Scalable flow-based networking with difane. SIGCOMM Comput. Commun. Rev. **41**(4) (2010)
10. K. Zheng, L. Wang, B. Yang, Y. Sun, S. Uhlig, Lazyctrl: A scalable hybrid network control plane design for cloud data centers. IEEE Trans Parallel Distrib Syst **99**, 1–1 (2016)
11. H. Mekky, F. Hao, S. Mukherjee, Z.L. Zhang, T. Lakshman, Application-aware data plane processing in sdn, in *Proceedings of the Third Workshop on Hot Topics in Software Defined Networking, HotSDN'14*, (ACM, New York, 2014), pp. 13–18
12. SDN Versus Traditional Networking Explained: https://www.cisco.com/c/dam/en_us/solutions/industries/docs/gov/software_defined_networking.pdf
13. Software-Defined Networking and Network Programmability: Use Cases for Defense and Intelligence Communities. https://www.ibm.com/services/network/sdn-versus-traditional-networking
14. K.-T. Foerster, S. Schmid, S. Vissiccho, Survey of consistent software-defined network updates. IEEE Commun Surv Tutor **21**, 1435–1461 (2018)
15. B.A.A. Nunes, M. Mendonca, X.N. Nguyen, K. Obraczka, T. Turltti, A survey of software-defined networking: Past, present, and future of programmable networks. IEEE Commun Surv Tutor **16**(3), 1617–1634 (2014)
16. S. Yeganeh, A. Tootoonchian, Y. Ganjali, On scalability of software-defined networking. IEEE Commun. Mag. **51**(2), 136–141 (2013). https://doi.org/10.1109/MCOM.2013.6461198
17. M. Blose, L.A. Akinyemi, S. Ojo, M. Faheem, A.L. Imoize, A.A. Khan, Scalable hybrid switching-driven software defined Networking issue: From the perspective of reinforcement learning standpoint. IEEE Access **12**, 63334–63350 (2024)
18. A. Voellmy, J. Wang, Y.R. Yang, B. Ford, P. Hudak, Maple: simplifying SDN programming using algorithmic policies, in *Proceedings of the ACM SIGCOMM 2013 conference on SIGCOMM, ser. SIGCOMM'13*, (ACM, New York, 2013), pp. 87–98
19. A. Tootoonchian, S. Gorbunov, Y. Ganjali, M. Casado, R. Sherwood, On controller performance in software-defined networks, in *Proceedings of the 2nd USENIX conference on Hot Topics in Management of Internet, Cloud, and Enterprise Networks and Services, ser. Hot-ICE'12*, (USENIX Association, Berkeley, 2012), pp. 10–10
20. T. Koponen, M. Casado, N. Gude, J. Stribling, L. Poutievski, M. Zhu, R. Ramanathan, Y. Iwata, H. Inoue, T. Hama, S. Shenker, Onix: a distributed control platform for large-scale production networks, in *Proceedings of the 9th USENIX conference on Operating systems design and implementation, ser. OSDI'10*, (USENIX Association, Berkeley, 2010), pp. 1–6
21. Z. Cai, A.L. Cox, T.S.E. Ng, *Maestro: A System for Scalable OpenFlow Control* (Rice University, Tech. Rep., 2011)
22. J.C. Mogul, P. Congdon, Hey, you darned counters!: Get off myasic! in *Proceedings of the First Workshop on Hot Topics in Software Defined Networks, ser. HotSDN'12*, (ACM, New York, 2012), pp. 25–30
23. C. Wang, L. Zhang, Z. Li, C. Jiang, Sdcor: Software-defined cognitive routing for the internet of vehicles. IEEE Internet Things J. **5**(5), 3513–3520 (2018)
24. D. Gopi, S. Cheng, R. Huck, Comparative analysis of SDN and conventional networks using routing protocols. In CITS. IEEE, 108–112 (2017)
25. A. Shirmarz, A. Ghaffari, An adaptive greedy flow routing algorithm for performance improvement in a software-defined network. Int. J. Numer. Modell. Electron. Networks Devices Fields **33**(1) (2020)
26. I. Martın, S. Troia, J.A. Hernández, A. Rodríguez, F. Musumeci, G. Maier, R. Alvizu, Ó.G. de Dios, Machine learning-based routing and wavelength assignment in software-defined optical networks. IEEE Trans. Netw. Serv. Manag. **16**(3), 871–883 (2019)
27. X. Xu, L. Zuo, Z. Huang, Reinforcement learning algorithms with function approximation: Recent advances and applications. Inf. Sci. **261**, 1–33 (2014)
28. S. S. Richard, G. B. Andrew, Reinforcement Learning. ebook. An Introduction Second Edition, pp. 257–284, 459–464 (2018)

29. M. Blose, L.A. Akinyemi, Development of Scalable Hybrid Switching Based Software Defined Networking, in *2020 23rd international symposium on wireless personal multimedia communications (WPMC2020)*, (Tokyo, IEEE, 2020), pp. 1–5
30. M. Blose, L.A. Akinyemi, F. Nicolls, N. Ventura, Development of scalable hybrid switching based SDN using reinforcement learning, in *Proceedings Southern Africa Telecommunication Networks Applications Conference (SATNAC), KwaZulu-Natal, South Africa*, vol. 29, (2023), pp. 163–168
31. M. Blose, L.A. Akinyemi, Scalable hybrid switching-inspired Sofware defined Networking challenges: From the perspective of machine learning approach, in *2024 Conference on Information Communications Technology and Society (ICTAS)*, (IEEE, 2024), pp. 186–191

# Machine Learning Algorithms for Energy Consumption Prediction in Smart Homes: A Comparative Study

**Donatien Koulla Moulla, David Attipoe, Lateef Adesola Akinyemi, Sree Ganesh Thottempudi, Jelil Olatunbosun Agbo-Ajala, Ernest Mnkandla, and Alain Abran**

## 1 Introduction

The rapid advancement of Internet of Things (IoT) technologies has significantly enhanced the capabilities of smart homes. Equipped with various sensors and devices, smart homes can monitor and control different aspects of the living environment, including energy consumption for lighting, heating, and cooling systems. Accurate energy consumption prediction is important for optimizing energy usage, reducing costs, and minimizing environmental impact. According to statistics from several organizations, including the World Energy Organization, the household and commercial sectors account for roughly 30% of worldwide energy consumption [1]. The concept of smart homes has gained significant traction in recent years, driven by the growing demand for energy efficiency and sustainability. Machine learning (ML) algorithms have emerged as powerful tools for energy prediction in smart homes, leveraging the vast amounts of data generated by smart home sensors and devices [2].

Research on smart homes and energy prediction has shown potential for cost and energy savings [3, 4]. Chen et al. [5] emphasized the need for monitoring,

D. K. Moulla (✉) · D. Attipoe · L. A. Akinyemi · S. G. Thottempudi · J. O. Agbo-Ajala
E. Mnkandla
Centre for Augmented Intelligence and Data Science, School of Computing, University of South Africa, Johannesburg, South Africa
e-mail: moulldk@unisa.ac.za; attipds@unisa.ac.za; akinyla@unisa.ac.za; sreeganesh.thottempudi@srh.de; agboaj@unisa.ac.za; mnkane@unisa.ac.za

A. Abran
Department of Software Engineering and Information Technology, École de Technologie Supérieure, Montréal, QC, Canada
e-mail: alain.abran@etsmtl.ca



I. Woungang, S. K. Dhurandher (eds.), *The 7th International Conference on Wireless, Intelligent and Distributed Environment for Communication*, Lecture Notes on Data Engineering and Communications Technologies 237,
https://doi.org/10.1007/978-3-031-80817-3_5

analyzing, and predicting energy usage in smart homes. Numerous studies have underscored the importance of using ML models for energy prediction in smart homes, highlighting their potential to accurately predict and guide energy use [6–11].

There are several challenges related to predicting energy consumption using ML models, including but not limited to the size and quality of the dataset collected (e.g., missing data, outliers), the choice of features (e.g., weather conditions, occupancy patterns, appliance usage), resource constraints, the selection of suitable models, and the effective interpretation of the results. More specifically, enhancing the accuracy and applicability of energy consumption prediction models in smart homes remains a challenge due to the diverse types of energy usage patterns [11].

This study aims to evaluate and compare the performance of various ML models in the context of energy prediction for smart homes to identify the most suitable models for practical implementation. Six ML models were considered in this study including LSTM networks, GRU, RF, SVM, ANN, and XGBoost. This comparative analysis enriches insights into the predictive capabilities of each model.

## *1.1 Main Contribution*

The primary objective of this study is to improve the accuracy of energy consumption prediction in smart homes by leveraging advanced ML models. The main contributions of this work are as follows:

- **Comprehensive evaluation of ML models:** This study presents a thorough evaluation of six leading ML algorithms for energy consumption prediction: LSTM, RF, XGBoost, GRU, SVM, and ANN. The models were assessed using key performance metrics including MSE, MAE, MAPE, and $R^2$.
- **Incorporation of weather data:** The study further investigates the impact of incorporating temperature data on the performance of the prediction models. The inclusion of weather data significantly improved the accuracy of the models, highlighting the importance of external factors in energy consumption forecasting.
- **Insights into model performance:** By evaluating the strengths and weaknesses of each model, the study provides valuable insights into their applicability for smart home energy management. The RF model was identified as the most robust and accurate model for this application, both with and without the inclusion of weather data.

The remainder of this paper is organized as follows. Section 2 presents an overview of related work on ML models for energy consumption prediction in smart homes. In Sect. 3, the proposed methodology is presented. Section 4 presents and discusses the results of the performance evaluation. Section 5 presents the conclusion, including a summary of key findings and directions for future work.

## 2 Related Work

Researchers have proposed various predictive ML models for energy consumption and explored their potential applications in optimizing energy consumption in smart homes. The study referenced in [12] developed a cohesive model for predicting energy consumption in smart homes using various algorithms, with the RF model outperforming others. Lin et al. [13] used data mining and optimization techniques to estimate energy consumption in smart homes, highlighting the importance of accurate heating and cooling energy predictions. The authors used SVM and RF models, which demonstrated superior efficiency over traditional methods.

Mathumitha et al. [14] reviewed deep learning techniques for smart building energy consumption forecasting. The authors identified challenges, suggested a hybrid model for enhanced prediction accuracy, and highlighted the impact of factors such as weather, occupant behavior, and load fluctuation on forecasting. Bourhnane et al. [7] used ANN and genetic algorithms (GAs) to predict and schedule energy consumption in smart buildings. They trained models using data from photovoltaic (PV) installations and electrical appliances, but faced limitations, including the limited dataset size and the focus on short-term predictions [15].

The study in [16] used ML models, including RF, decision tree (DT), XGBoost, and AdaBoost, to predict hourly energy consumption in South African residential buildings, revealing that RF and DT models outperformed boosting models. In another study, Geetha et al. [17] proposed the use of a RF supervised learning model to predict domestic power peak demand and consumption. The study found that RF outperformed other models in terms of accuracy, stability, and generalization. The residential sector's total electricity demand is dominated by the energy consumption of appliances [18].

Malik et al. [19] used deep learning models, including ARIMA, SARIMAX, LSTM Univariate, and LSTM Multivariate, to improve energy consumption forecasting accuracy in smart homes. The study in [20] investigated the use of ML algorithms to optimize energy consumption in smart homes, using LR, GBoostR, DTR, SGDR, and BRR. However, the study was limited by the scope of the dataset used, which may not cover all scenarios in smart home energy consumption.

Syamala et al. [21] explored the use of deep learning and IoT in predicting energy consumption in smart residential buildings, highlighting the significance of optimal window size for prediction performance and model uncertainty estimation. They highlighted the economic impact of IoT-based technologies and the potential for smart home systems to reduce energy usage and forecast future demand.

A related study [22] proposed a model to predict solar radiation and energy consumption for smart homes using ML techniques, such as gradient boosting and SVM. A comparative study of different ML models was conducted to determine the best method for building a prediction model, and the results showed deep neural networks provided the highest accuracy.

Ozdemir [23] investigated the use of AI and ML techniques to forecast Turkey's long-term electrical energy demand from 1975 to 2020. The research analyzed historical consumption and socioeconomic indicators, using metaheuristic algorithms and AI approaches to provide reliable energy plans for sustainable development. The authors in [24] compared various machine learning models for predicting electrical energy consumption in smart homes, finding XGBoost with the lowest RMSE value of 0.318.

Khaoula et al. [25] evaluated three energy consumption prediction models: linear regression, RF, and ANN, using data from a smart home case study. The results showed that the ANN model had the highest accuracy. The study in [26] advocates using LSTM for energy consumption forecasting with realistic smart meter data. The results demonstrated that the LSTM approach achieves high accuracy in predicting energy consumption by modeling long temporal sequences, benefiting smart grid decision-making. Durand et al. [27] analyzed energy consumption variables in smart buildings, using techniques like Pearson and Spearman correlations, multiple linear regression, PCA, autocorrelation, and ARIMA. They found LSTM effective in handling time-series data for energy consumption prediction.

In summary, several ML models have been proposed for predicting energy consumption in smart homes. RF consistently outperforms other models in accuracy, stability, and generalization, though its generalization to other regions or time periods remains unexplored. While models such as SVM and deep learning techniques show efficiency and accuracy, their applicability is often limited by dataset size, quality, or specific conditions, such as geographic data. Several studies emphasize the need for hybrid models and improved data handling, with challenges including scalability, integration, and managing diverse IoT data. Overall, the inclusion of external factors such as weather data can significantly enhance model performance.

## 3 Proposed Methodology

This section discusses the datasets used for this study. It provides a brief description of the datasets, including all the features considered. Furthermore, it examines the prediction framework adopted to help the ML algorithms perform optimally.

### *3.1 Dataset Characterization*

#### 3.1.1 Electric Power Consumption

This study employs a comprehensive dataset of electric power consumption measurements from a single household, recorded every minute over nearly 4 years, from December 2006 to November 2010. The dataset includes various electrical quantities and sub-metering values, totaling 2,075,259 instances and 9 features. It is particularly suitable for tasks such as regression and clustering, given its multivariate and time-series characteristics. The data were collected from a house in Sceaux, located 7 km from Paris, France. The dataset's features include timestamp, global active power, global reactive power, voltage, global intensity, and sub-metering values for the kitchen, laundry room, and climate control system. These features provide detailed information on the household's power consumption patterns. Notably, the active energy consumed by unmeasured electrical equipment can be calculated using the following formula:

$$\text{Global avtive power} \times 1000/60 - \text{Submetering}_1 - \text{Submetering}_2 - \text{Submetering}_3 \quad (1)$$

The dataset contains some missing values, approximately 1.25% of the total, represented by the absence of data between two consecutive semicolons. An example of this can be seen on April 28, 2007 [25, 26]. To address this, the repository authors preprocessed the dataset by removing rows with significant missing data and performed interpolation where feasible. Additionally, to facilitate combined analysis with an hourly dataset from another source, the daily data is granularized by averaging the values to match the hourly timestamps. This dataset, sourced from the UC Irvine Machine Learning Repository (MLR), serves as an ideal foundation for developing and testing various predictive models, such as RF, XGBoost, ANN, SVM, LSTM, and GNU, aimed at analyzing and forecasting power consumption. More information about the dataset can be found in [25, 27]. Figure 1 presents the electric power consumption dataset from December 2006 to November 2010. As discussed in the Introduction section, models like LSTM usually require large datasets to improve their prediction accuracy. To this end, we introduce comprehensive weather data from Sceaux for the same period.

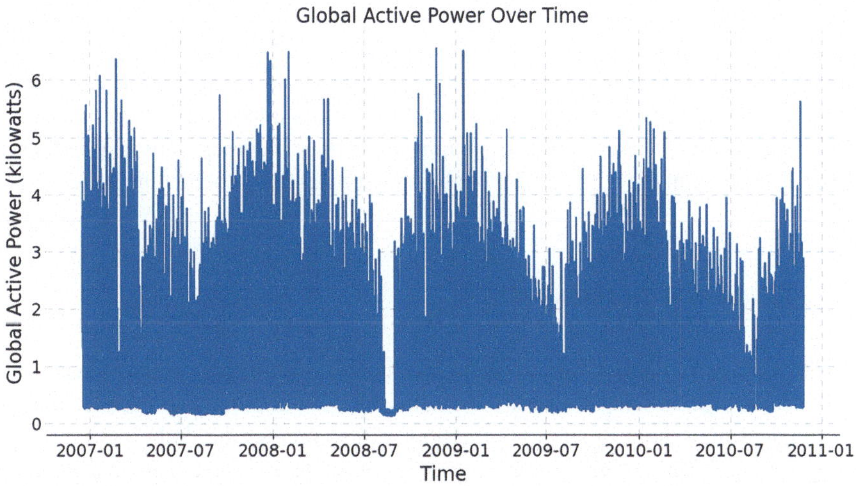


**Fig. 1** Electric power consumption dataset from December 2006 to November 2010

**Table 1** Summary of the power consumption dataset

| Statistics | GAP | GRP | Voltage | GI | $SubM1$ | $SubM2$ | $SubM3$ |
|---|---|---|---|---|---|---|---|
| Count | 34168 | 34168 | 34168 | 34168 | 34168 | 34168 | 34168 |
| Mean | 1.091 | 0.124 | 240.839 | 4.628 | 1.122 | 1.299 | 6.459 |
| Std | 0.898 | 0.067 | 3.001 | 3.758 | 3.538 | 4.193 | 7.351 |
| Minimum | 0.124 | 0 | 225.835 | 0.503 | 0 | 0 | 0 |
| 25% | 0.342 | 0.078 | 239.181 | 1.513 | 0 | 0 | 0.65 |
| 50% | 0.803 | 0.107 | 240.978 | 3.423 | 0 | 0.333 | 1.65 |
| 75% | 1.579 | 0.149 | 242.695 | 6.607 | 0 | 0.65 | 14.217 |
| Maximum | 6.561 | 0.774 | 251.902 | 28.383 | 48.367 | 46.433 | 21.55 |

GAP means global active power, GRP means global reactive power, GI means global intensity, and SubM means submetering

### 3.1.2 Weather Dataset Description

The weather dataset used in this study was obtained from Visual Crossing, a comprehensive repository of historical weather data. This dataset contains daily weather observations for Sceaux (7 km), which span the same time-period as the household power consumption data, from December 2006 to November 2010. The dataset contains a variety of meteorological characteristics, like "Temperature," which captures the maximum, minimum, and average temperatures as *tempmax, tempmin, temp*. It also captures "Feels Like" of maximum, minimum, and average feels-like temperatures as *feelslikemax, feelslikemin* and *feelslike*. Other features of the dataset include "Dew Point" as dew, "Humidity" as humidity, "Precipitation" as *precip, precipprob, precipcover* and *preciptype*, "Snow" as *snow* and *snowdepth*, and "Wind" as *windspeed, windgust* and *winddir*, representing speed, gust, and direction of the wind, respectively. Indeed, there are 18 columns in total with other meteorological markers like sea-level pressure, visibility, cloud cover, solar radiation and solar energy, UV index, sunrise and sunset times, and moon phase. This detailed weather data provides essential contextual information that can impact household power consumption, thus improving the accuracy of the predictive models. The summary of the power consumption dataset is provided in Table 1.

## *3.2 Prediction Framework*

The prediction framework for this study is designed to accurately predict electricity power consumption using a variety of ML algorithms. Our initial approach leverages the detailed household power consumption data to predict the daily and monthly consumption. Furthermore, we incorporate the weather data to enhance the prediction capability of more advanced models like LSTM and GRU.

### 3.2.1 Preprocessing (Data Cleaning and Integration)

Following from the already extensive preprocessing work done by Lichman [28], we go through the following steps to ensure the dataset is aligned with our study:

- **Data Cleaning:** In this instance, both the power consumption dataset and the weather dataset were cleaned to handle missing values. In the power consumption dataset, rows with significant missing data were removed. For the weather data, all categorical date columns such as PrecipType were dropped, and rows with irrelevant timestamps were removed to ensure consistency between the datasets. Furthermore, the weather data had duplicate datetime entries, which were identified and removed to ensure each timestamp was unique.
- **Data Integration:** The weather data, originally at a daily granularity, was aligned with the power consumption data, which is at a 1-hour resolution. The daily weather data was granularized to match the hourly format of the power consumption data by averaging the values for each hour.

As previously mentioned, various ML regression and autoregressive models can be used to predict power consumption based on past values. However, there are several exogenous variables that can affect usage, thereby influencing the performance of the ML models. To demonstrate this effect, we analyze the daily load profiles of the time series in juxtaposition with these exogenous factors. Figure 2 demonstrates the variation in daily load profiles given certain calendar features. All figures from Figs. 2, 3, and 4 illustrate the average daily load of the dataset as a reference point, while they additionally depict: (i) the average daily load on holidays versus

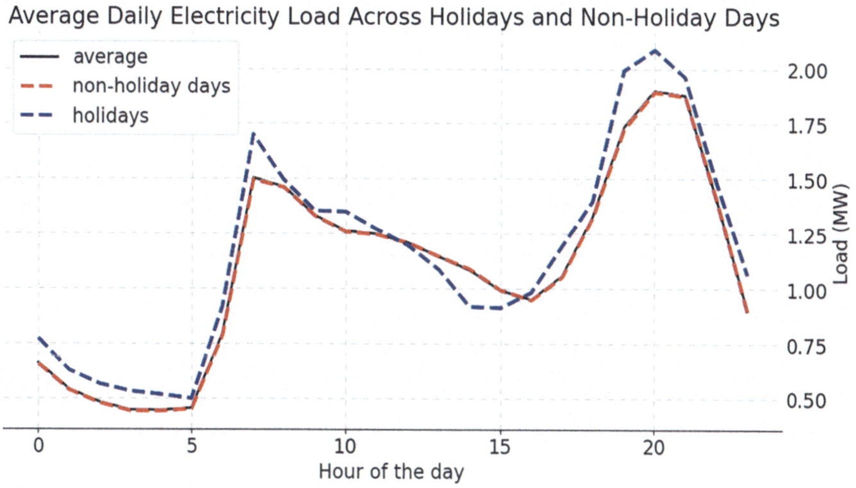


**Fig. 2** Average daily electricity load across holidays and non-holiday days

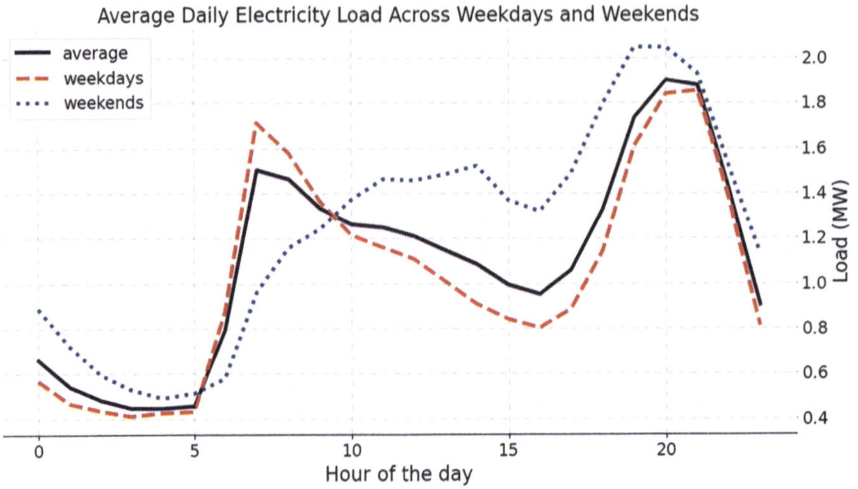


**Fig. 3** Average daily electricity load across weekdays and weekend days

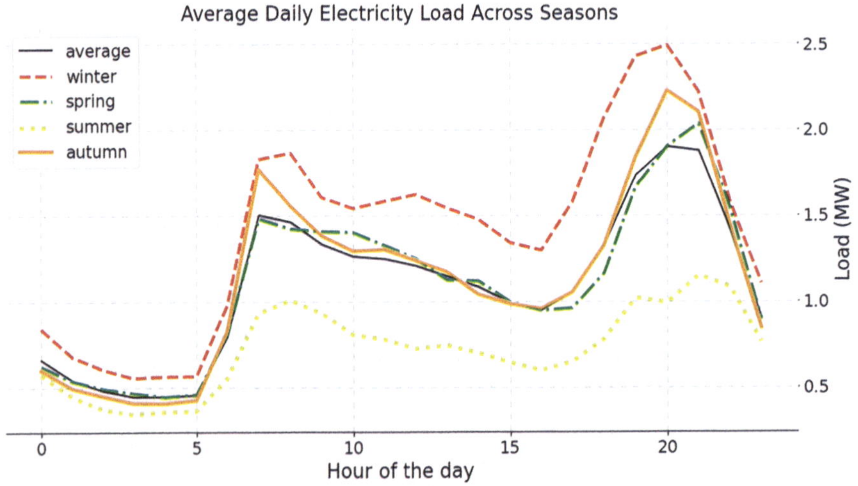


**Fig. 4** Average daily electricity load across various seasons

non-holiday days, (ii) the average daily load on weekends versus weekdays, and (iii) the inter-season daily profiles. We did not consider temperature profiles in this case for the sake of brevity

Indeed, from Fig. 2, we see how the electricity consumption patterns differ between holidays and non-holidays. On holidays, the load peaks in the late morning and evening as people spend more time at home as expected. For Fig. 3, the consumption patterns differ between weekdays and weekends. Weekdays exhibit higher early morning and evening loads, while weekends have lower early morning loads but peak in the late morning and early afternoon due to different activity schedules. As we scale up to look at the seasonality in the data, we can see in Fig. 4 that the consumption varies significantly with seasons, influenced by heating and cooling needs. Winter and autumn show higher electricity usage, particularly in the evening, while summer shows a lower load during the day with an evening peak.

#### 3.2.2 Feature Extraction and Engineering

We ended up working with two datasets, the power consumption data without weather parameters and the combined dataset with weather parameters. The combined dataset includes a range of features such as temperature, humidity, wind speed, and various meteorological measurements. These features were categorized into four groups: temperature (e.g., tempmax, tempmin, feelslike), precipitation (e.g., precip), wind (e.g., windspeed, windgust), and atmospheric conditions (e.g., humidity, pressure).

Furthermore, feature scaling was performed to normalize the data, making it suitable for ML algorithms. This step involved transforming the data to a common scale without distorting differences in the ranges of values. Additional temporal features were extracted from the data to ensure all the underlying seasonality of the data is properly captured.

### *3.3 Machine Learning Models*

Our prediction framework involves six ML algorithms: RF, XGBoost, ANN, SVM, LSTM, and GRU. Each model was trained on the cleaned dataset to predict the household energy consumption. The detailed description of each model is provided in the following sections.

#### 3.3.1 Random Forest (RF)

Random forest (RF) is an ensemble learning model that builds multiple decision trees and combines their predictions to improve accuracy and reduce overfitting. It operates by constructing multiple trees during training and outputting the mean prediction of the individual trees for regression tasks. Mathematically, for a given input vector $\mathbf{x}$, the prediction $\hat{y}$ of the random forest can be expressed as:

$$\hat{y} = \frac{1}{N} \sum_{i=1}^{N} \mathcal{T}_i(\mathbf{x}) \tag{2}$$

where $\mathcal{T}_i(x)$ is the prediction of the $i$-th tree and $N$ is the total number of trees.

### 3.3.2 Extreme Gradient Boosting (XGBoost)

Extreme gradient boosting is a scalable and efficient gradient boosting algorithm that builds models sequentially to correct errors made by previous models. It uses a second-order Taylor approximation for optimization, which improves accuracy and performance [29]. The objective function for XGBoost is given by:

$$\mathcal{L}(\theta) = \sum_{i=1}^{n} l(y_i, \hat{y}_i) + \sum_{k=1}^{K} \Omega(f_k), \tag{3}$$

where $l$ is a a differentiable convex loss function that measures the difference between the prediction $\hat{y}_i$ and the target $y_i$, and $\Omega$ is a regularization term that penalizes the complexity of the model to prevent overfitting. The prediction $\hat{y}_i$ is the sum of the predictions from each tree:

$$\hat{y}_i = \sum_{k=1}^{K} f_k(\mathbf{x}_i).$$

### 3.3.3 Artificial Neural Network (ANN)

ANNs consist of multiple layers of interconnected nodes (neurons) that transform the input data through a series of weights and activation functions. For our purposes, we used a multilayer perceptron (MLP) with multiple hidden layers. The mathematical formulation for an ANN with one hidden layer is:

$$\mathbf{h} = \sigma\left(\mathbf{W}_h\mathbf{x} + \mathbf{b}_h\right)$$
$$\hat{y} = \mathbf{W}_0\mathbf{h} + \mathbf{b}_0$$

where $\mathbf{x}$ is the input vector and $\mathbf{W}_h$ and $\mathbf{b}_h$ are the weights and biases of the hidden layer. $\mathbf{W}_0$ and $\mathbf{b}_0$ are the weights and biases of the output layer, and $\sigma$ is an activation function with the rectified linear unit (ReLU) being the most popular.

### 3.3.4 Support Vector Machine (SVM)

Support vector machine (SVM) is a supervised learning algorithm that finds the optimal hyperplane for regression tasks by using the $\epsilon$-insensitive loss function, which ignores errors within a certain margin. The objective function for SVM regression is:

$$
\begin{aligned}
\min_{w,b,\xi,\xi^*} \frac{1}{2}\|\mathbf{w}\|^2 + C\sum_{i=1}^{n}\left(\xi_i + \xi_i^*\right), \\
\text{subject to } y_i - (\mathrm{w}\cdot \mathrm{x}_i + b) \leq \epsilon + \xi_i \\
(\mathrm{w}\cdot \mathrm{x}_i + b) \leq \epsilon + \xi_i \\
\xi, \xi^* \geq 0,
\end{aligned}
\tag{4}
$$

where $\mathbf{W}$ is a weight vector, $b$ is the bias, $\xi$ and $\xi^*$ are slack variables, $C$ is a regularization parameter, and $\epsilon$ is the margin of tolerance.

### 3.3.5 Long Short-Term Memory (LSTM)

Long short-term memory (LSTM) networks, a type of recurrent neural network (RNN), are designed to handle sequential data and capture long-term dependencies by introducing a memory cell and three different gates, that is, input gate, output gate, and forget gate. The cells essentially maintain a hidden state over time and the gates control the flow of information. The architecture for the LSTM cell are mathematically represented by:

$$
\begin{aligned}
f_t &= \sigma\left(\mathbf{W}_f \cdot [h_{t-1}, x_t] + b_f\right), \\
i_t &= \sigma\left(\mathbf{W}_i \cdot [h_{t-1}, x_t] + b_i\right), \\
o_t &= \sigma\left(\mathbf{W}_o \cdot [h_{t-1}, x_t] + b_o\right), \\
g_t &= \tanh\left(\mathbf{W}_g \cdot [h_{t-1}, x_t] + b_g\right), \\
c_t &= f_t \cdot c_{t-1} + i_t \cdot g_t, \\
h_t &= 0_t \cdot \tanh(c_t),
\end{aligned}
\tag{5}
$$

where $f_t, i_t, o_t$ are the forget, input, and output gates, $g_t$ is the cell state, $c_t$ is the memory cell, $h_t$ is the hidden state, $\xi$ and $\xi^*$ are slack variables, $C$ is a regularization parameter, $h_t$ is the hidden state at time step $t$, and $b_k$ is the respective bias at gate $k$. In general, backpropagation through time is used to train calculate the gradients of the architecture while gradually updating their weights.

### 3.3.6 Gated Recurrent Unit (GRU)

The gated recurrent unit (GRU) is a type of RNN designed to address the vanishing gradient problem in standard RNNs. GRUs, like long short-term memory (LSTM) units, are particularly well-suited for sequential data and time-series forecasting tasks. However, they have fewer gates and are known to be computationally more efficient than LSTMs while still capturing long-term dependencies.

Unlike the LSTM, a GRU has two main gates: the update gate and the reset gate. These gates control the flow of information and help the network decide what information to keep and what to discard. In essence, the GRU is governed by:

$$\begin{aligned} z_t &= \sigma\left(\mathbf{W}_z \cdot [h_{t-1}, x_t]\right), \\ r_t &= \sigma\left(\mathbf{W}_r \cdot [h_{t-1}, x_t]\right), \\ \hat{h}_t &= \tanh\left(\mathbf{W} \cdot [r_t \cdot h_{t-1}, x_t]\right), \\ h_t &= (1 - z_t) \cdot h_{t-1} + z_t \cdot \hat{h}_t, \end{aligned} \tag{6}$$

where $z_t, r_t$ are the update and reset gates, respectively, controlling how much of the past information to carry forward or forget. $\hat{h}_t$ is the candidate hidden state and $h_t$ is the final hidden state.

## 3.4 Setting of the Study

In this study, the ML models were trained using features extracted from the UC Irvine Machine Learning Repository dataset [28] and weather dataset utilizing Python libraries. These comprehensive datasets include various electrical quantities and sub-metering values, as well as temperature and other meteorological values, providing a robust foundation for analyzing household energy consumption patterns.

The combined dataset was split into training and testing sets to evaluate the performance of the models. The training set consisted of data from December 2006 to October 2009, while the testing set comprised data from November 2009 to November 2010. This split ensures that the models are trained on a substantial portion of the data while retaining a significant period for testing and validation.

Table 2 provides the information on the environment used in this study.

**Table 2** The setting environment of the study

| Component | Description |
|---|---|
| Location | Sceaux, Île-de-France, France |
| Period | December 2006–November 2010 |
| Data sources | UC Irvine Machine Learning Repository |
| Key features | Time, global active power, global reactive power, voltage, global intensity, sub-metering 1, sub-metering 2, sub-metering 3, temperature, precipitation |
| Train/test split | Training, December 2006–October 2009; testing, November 2009–November 2010 |
| Tools | Python, scikit-learn, Pandas, Numpy, Matplotlib, Holidays library, Visual Crossing weather data |
| Hardware | Standard computing hardware (e.g., Intel Core i7 processor, 16 GB RAM) Software Python 3.8, Jupyter Notebook, SciPy stack |

### 3.5 Performance Evaluation Metrics

This section presents the evaluation criteria used to evaluate the performance of the chosen models: the mean squared error (MSE), mean absolute error (MAE), mean absolute percentage error (MAPE), and coefficient of determination ($R^2$). The models are defined below:

- Mean Absolute Error (MAE): MAE measures the average absolute deviation between the predicted and actual values, giving all errors equal weighting. It is widely used in machine learning for its straightforward interpretation:

$$\mathrm{MAE} = \frac{\sum_{i=1}^{n} \left| y_i - \hat{y}_i \right|}{n}. \tag{7}$$

- Mean Squared Error (MSE): MSE measures the average scale of the squared differences between the predicted and actual values. It penalizes larger errors more heavily than smaller ones, providing a sensitive measure of prediction accuracy:

$$\mathrm{MSE} = \frac{1}{n} \sum_{i=1}^{n} \left( y_i - \hat{y}_i \right)^2, \tag{8}$$

- Mean Absolute Percentage Error (MAPE): MAPE measures the accuracy of the predictions as a percentage. It is particularly useful for comparing the prediction accuracy across different scales:

$$\mathrm{MAPE} = \frac{1}{n} \sum_{i=1}^{n} \frac{\left| y_i - \hat{y}_i \right|}{y_i}, \tag{9}$$

- Coefficient of Determination ($R^2$): $R^2$ indicates the proportion of the variance in the dependent variable that is predictable from the independent variables. It provides an indication of the goodness of fit of the model:

$$\mathrm{R}^2 = 1 - \frac{\sum_{i=1}^{n} \left( y_i - \hat{y}_i \right)^2}{\sum_{i=1}^{n} \left( y_i - \overline{y}_i \right)^2}, \tag{10}$$

where $y_i$ is the actual measurement, $\hat{y}_i$ is the predicted value, $\overline{y}_i$ is the mean of the actual target values, and $n$ is the number of measurements. These metrics were chosen to ensure a comprehensive evaluation of model performance, considering both absolute and relative errors, as well as the overall fit of the model to the data.

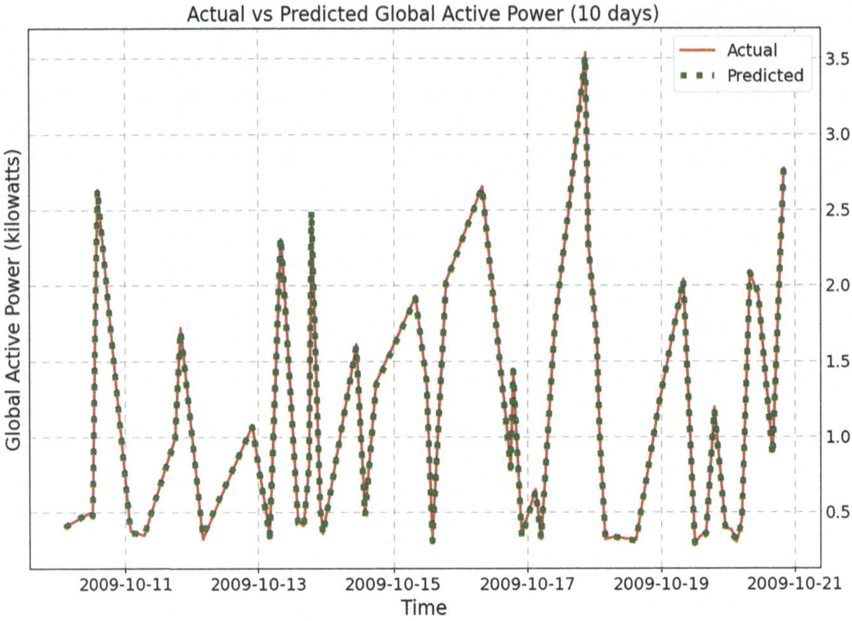


**Fig. 5** Power consumption data prediction for 10 days (October 10–20, 2009)

# 4 Results and Discussions

In this section, the results of the predictive models are presented and discussed. Firstly, we present the initial prediction results with only the power consumption dataset. Figure 5 represents the result from running the model with just the power consumption data.

## *4.1 Model Performance Without Weather Data*

The performance of various ML models for power consumption prediction without incorporating weather data is summarized in Table 3. The key metrics used for evaluation include mean absolute error (MAE), mean squared error (MSE), the coefficient of determination ($R^2$), and mean absolute percentage error (MAPE).

- **Random Forest**: This model exhibited the best overall performance with the lowest MAE (0.011290) and MSE (0.000339) and the highest $R^2$ (0.999567). The MAPE (0.015335) also indicates a very accurate prediction, demonstrating the robustness of RF in capturing the underlying patterns of the data.

**Table 3** Performance of the prediction models for power consumption dataset without weather data

| Model | MAE | MSE | $R^2$ | MAPE |
|---|---|---|---|---|
| Random forest | 0.011290 | 0.000339 | 0.999567 | 0.015335 |
| XGBoost | 0.012549 | 0.000764 | 0.999024 | 0.015977 |
| SVM | 0.041845 | 0.004363 | 0.994424 | 0.063269 |
| ANN | 0.012776 | 0.000335 | 0.999572 | 0.020175 |
| LSTM | 0.015744 | 0.000426 | 0.999456 | 0.022648 |
| GRU | 0.012862 | 0.000329 | 0.999580 | 0.018114 |

- **XGBoost**: Indeed, XGBoost also performed well, with slightly higher MAE (0.012549) and MSE (0.000764) compared to RF. Again, the $R^2$ value (0.999024) suggests that it explains most of the variance in the data. The MAPE (0.015977) is marginally higher than random forest.
- **SVM**: The SVM model showed reasonable performance but with higher errors (MAE, 0.041845; MSE, 0.004363) and a lower $R^2$ (0.994424). The MAPE (0.063269) was the highest among the models, indicating that SVM struggled more with accurate predictions.
- **ANN**: The ANN model performed well, with an MAE of 0.012776 and MSE of 0.000335, closely matching the performance of random forest. The $R^2$ value (0.999572) was high, and the MAPE (0.021075) was reasonable, making ANN a reliable choice for prediction.
- **LSTM**: The model had higher MAE (0.015744) and MSE (0.000426) compared to random forest and ANN. However, the $R^2$ value (0.999456) was still high, indicating good overall performance. The MAPE (0.022648) was higher than ANN but lower than SVM.
- **GRU**: In general, the GRU model performed similarly to the LSTM, with an MAE of 0.012862 and MSE of 0.000329. The $R^2$ (0.999580) was high, and the MAPE (0.018114) was lower than LSTM, making it a strong contender among the neural network models.

We present the graphs to show how each of the models performed in predicting the power consumption of the house. Figures 6 and 7 present the graphical representations of a 2-day prediction (actual vs predicted) of all the algorithms using the power consumption dataset only.

## 4.2 *Model Performance with Weather Data (Temperature Only)*

Incorporating weather data, specifically temperature, further improved the prediction accuracy, as shown in Table 4. Figure 8 presents the model prediction with power consumption data plus temperature only.

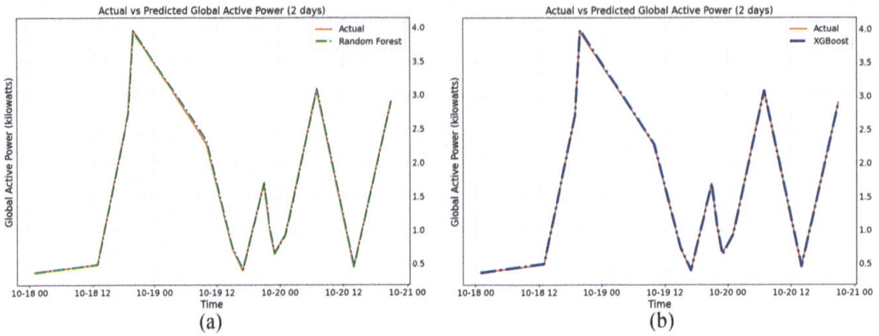


**Fig. 6** A 2-day prediction (actual vs predicted) of all the algorithms using the power consumption dataset only. (**a**) Random forest algorithm. (**b**) XGBoost algorithm

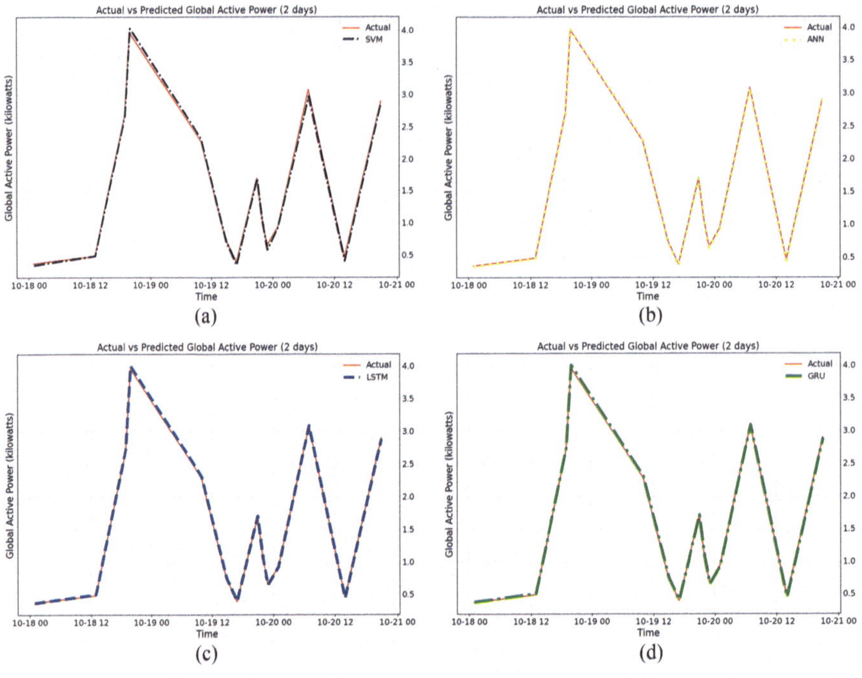


**Fig. 7** A 2-day prediction (actual vs predicted) of all the algorithms using the power consumption dataset only. (**a**) SVM algorithm. (**b**) ANN algorithm. (**c**) LSTM algorithm (**d**) GRU algorithm

- **Random Forest**: The inclusion of temperature data slightly improved the performance of the RF model, with an MAE of 0.011252 and MSE of 0.000315. The $R^2$ increased to 0.999606, and the MAPE remained consistent at 0.015336, confirming its robustness.

**Table 4** Performance of the prediction models for power consumption dataset with weather data (temperature only)

| Model | MAE | MSE | $R^2$ | MAPE |
|---|---|---|---|---|
| Random forest | 0.011252 | 0.000315 | 0.999606 | 0.015336 |
| XGBoost | 0.013010 | 0.001615 | 0.997978 | 0.015833 |
| SVM | 0.042781 | 0.004963 | 0.993787 | 0.072056 |
| ANN | 0.012036 | 0.000293 | 0.999634 | 0.019431 |
| LSTM | 0.012518 | 0.000318 | 0.999602 | 0.020138 |
| GRU | 0.011670 | 0.000275 | 0.999655 | 0.018691 |

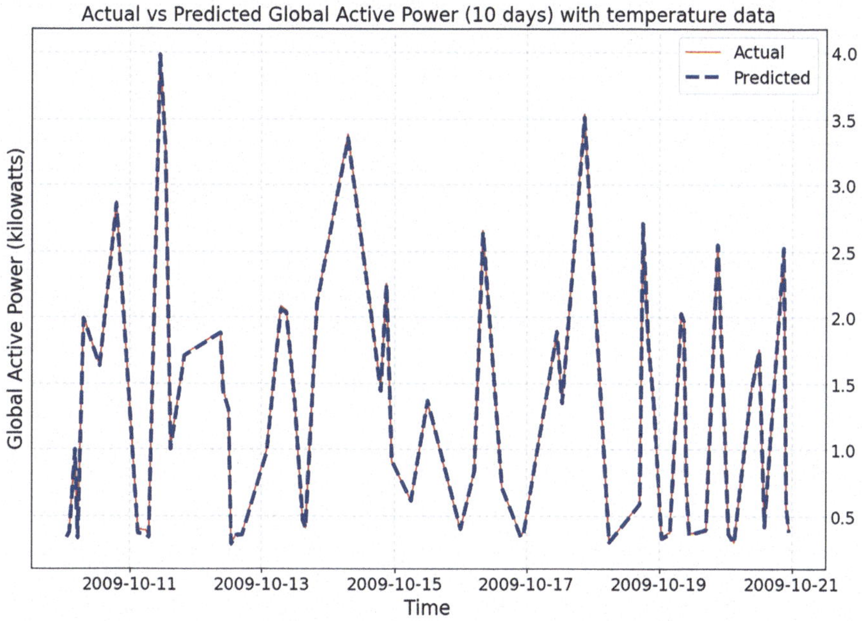


**Fig. 8** Model prediction with power consumption data plus temperature only

- **XGBoost**: XGBoost showed a noticeable improvement with weather data, achieving an MAE of 0.013010 and MSE of 0.000615. The $R^2$ increased to 0.999778, and the MAPE decreased to 0.015833, indicating enhanced accuracy.
- **SVM**: The SVM model's performance did not improve significantly with weather data, with an MAE of 0.042781 and MSE of 0.004963. The $R^2$ decreased to 0.993787, and the MAPE increased to 0.072056, suggesting that SVM did not benefit much from the additional feature.
- **ANN**: ANN showed improved performance with an MAE of 0.012036 and MSE of 0.000293. The $R^2$ increased to 0.999634, and the MAPE decreased to 0.019431, making it more accurate than without weather data.

- **LSTM**: The LSTM model's performance improved significantly with an MAE of 0.012518 and MSE of 0.000318. The $R^2$ increased to 0.999602, and the MAPE decreased to 0.020138, indicating better accuracy with temperature data.
- **GRU**: The GRU model also benefited from the inclusion of weather data, with an MAE of 0.011670 and MSE of 0.000275. The $R^2$ increased to 0.999655, and the MAPE decreased to 0.018691, making it the best performing neural network model with weather data.

Indeed, the inclusion of temperature data generally improved the performance of most models, highlighting the importance of weather data in predicting power consumption. This is particularly evident in the improved metrics for RF, XGBoost, ANN, LSTM, and GRU. The RF model consistently showing the best performance across all metrics, both with and without weather data, suggests its robustness as a model for power consumption prediction, capable of handling various features effectively. Furthermore, among the neural network models, GRU demonstrated the best performance with weather data, followed closely by ANN and LSTM. This indicates that GRU can capture temporal dependencies effectively when supplemented with additional features like temperature. Indeed, the observations here are in line with the work of other authors in the literature. As was observed, the SVM model did not perform as well as the other models, particularly with weather data. This may suggest that SVM might not be as effective for this type of time-series prediction, potentially due to its sensitivity to feature scaling and non-linearity.

## 5 Conclusion and Future Work

This study investigated several ML models to predict the power consumption of smart residential buildings using the UC Irvine Machine Learning Repository dataset. The models applied include random forest (RF), XGBoost, support vector machine (SVM), artificial neural network (ANN), long short-term memory (LSTM), and gated recurrent unit (GRU). Our findings indicate that the RF model consistently outperformed the other models in terms of prediction accuracy, showing the lowest MAE and MSE, and the highest $R^2$, both with and without incorporating weather data. The inclusion of temperature data improved the performance of most models, highlighting the importance of external factors in predicting power consumption. The ANN and GRU models also showed strong performance, particularly when weather data was included. The ANN performed well in capturing the underlying patterns in the data, while the GRU demonstrated its ability to handle temporal dependencies effectively. On the other hand, the SVM model did not perform as well as the others, particularly with the inclusion of weather data, suggesting that it may not be as suitable for this type of time-series prediction.

Future research will focus on hyperparameter tuning of the models to further enhance predictive accuracy and integrating appliance-level data for forecasting. This can be integrated into IoT systems for real-time monitoring and predictive maintenance, enabling better resource utilization and energy conservation.

## References


1. M. González-Torres, L. Pérez-Lombard, J.F. Coronel, I.R. Maestre, D. Yan, A review on buildings energy information: trends, end-uses, fuels and drivers. Energy Rep. **8**, 626–637 (2022)
2. M.S. Aliero, M. Asif, I. Ghani, M.F. Pasha, S.R. Jeong, Systematic review analysis on smart building: challenges and opportunities. Sustainability **14**(5), 3009 (2022)
3. M. Chammas, A. Makhoul, J. Demerjian, An efficient data model for energy prediction using wireless sensors. Comput. Electr. Eng. **76**, 249–257 (2019)
4. B. Aksanli, Accurate and data-limited prediction for smart home energy management, in *Proceedings of the ASME 2018 12th International Conference on Energy Sustainability collocated with the ASME 2018 Power Conference and the ASME 2018 Nuclear Forum. ASME 2018 12th International Conference on Energy Sustainability*. Lake Buena Vista, Florida, USA. June 24–28, 2018
5. C. Chen, B. Das, J. Cook, Diane, Energy prediction in smart environments, in *Ambient Intelligence and Smart Environments, Volume 8 Workshop Proceedings of the 6th International Conference on Intelligent Environments* (2010), pp. 148–157
6. X. Xiong, Y. Wei, The analysis and predication of energy use in smart homes based on machine learning, in *2020 International Conference on Computing and Data Science (CDS)*, Stanford (2020), pp. 381–386
7. S. Bourhnane, M.R. Abid, R. Lghoul et al., Machine learning for energy consumption prediction and scheduling in smart buildings. SN Appl. Sci. **2**(297), 1–10 (2020)
8. G. Revati, J. Hozefa, S. Shadab, A. Sheikh, S.R. Wagh, N.M. Singh, Smart building energy management: load profile prediction using machine learning, in *2021 29th Mediterranean Conference on Control and Automation (MED)*, Puglia, Italy (2021), pp. 380–385
9. O. Almughram, B. Zafar, S.B. Slama, Home energy management machine learning prediction algorithms: a review, in *Proceedings of the 2nd International Conference on Industry 4.0 and Artificial Intelligence (ICIAI 2021)* (Atlantis Press, 2022), pp. 40–47
10. N. Kumar, K. Sundaram, R. Reena, S. Madhumathi, Optimizing energy consumption in smart homes using machine learning techniques, in *E3S Web of Conferences*, vol. 387 (2023), pp. 1–7
11. S. Iram, H. Al-Aqrabi, H.M. Shakeel, H.M. Farid, M. Riaz, R. Hill, P. Vethathir, T.A Alsboui, An innovative machine learning technique for the prediction of weather based smart home energy consumption. IEEE Access **11**, 76300–76320 (2023)
12. S. Toshpulotova, M. Fayaz, A cohesive machine learning-based model for energy consumption prediction in smart homes, in *2022 International Conference on Electrical and Computing Technologies and Applications (ICECTA)*, Ras Al Khaimah, United Arab Emirates (2022), pp. 163–167
13. M. Lin, L. Peng, T. Liu, I. Zhang, Novel approach to energy consumption estimation in smart homes: application of data mining and optimization techniques. Front. Energy Res. **12**, 1–12 (2024)
14. R. Mathumitha, P. Rathika, K. Manimala, Intelligent deep learning techniques for energy consumption forecasting in smart buildings: a review. Artif. Intell. Rev. **57**(35), 1–33,2024 (2024)

15. X. Wang, H. Wang, B. Bhandari et al., AI-empowered methods for smart energy consumption: a review of load forecasting, anomaly detection and demand response. Int. J. Precis. Eng. Manuf.-Green Technol. **11**, 963–993 (2024)
16. D.K. Moulla, D. Attipoe, E. Mnkandla, A. Abran, Predictive model of energy consumption using machine learning: a case study of residential buildings in South Africa. Sustainability **16**(11), 4365 (2024)
17. R. Geetha, K. Ramyadevi, M. Balasubramanian, Prediction of domestic power peak demand and consumption using supervised machine learning with smart meter dataset. Multimedia Tools Appl. **80**, 19675–19693 (2021)
18. N. Arghira, L. Hawarah, S. Ploix, M. Jacomino, Prediction of appliances energy use in Smart homes. Energy **48**(1), 128–134 (2012)
19. S. Malik, S. Malik, I. Singh et al., Deep learning based predictive analysis of energy consumption for smart homes. Multimedia Tools Appl. 1–22 (2024)
20. M.A. Hameed, E.T. Yassen, W.M. Jasim, Enhancement methods for energy consumption prediction in smart house based on machine learning. Iraqi J. Comput. Sci. Math. **4**(4), 92–99 (2023)
21. S. Syamala et al., *Machine learning-integrated IoT-Based Smart Home Energy Management System. Handbook of Research on Deep Learning Techniques for Cloud-Based Industrial IoT*, ed. by P. Swarnalatha, S. Prabu (IGI Global, 2023), pp. 219–235
22. A. Dhage, A. Kakade, G. Nahar, M. Pingale, S. Sonawane, A. Ghotkar, Recommendation and prediction of solar energy consumption for smart homes using machine learning algorithms, in *2021 International Conference on Artificial Intelligence and Machine Vision (AIMV)*, Gandhinagar, India (2021), pp. 1–5
23. G. Ozdemir, Long-term electrical energy demand forecasting by using artificial intelligence/machine learning techniques. Electr. Eng. **106**, 5229–5251 (2024)
24. M.H. Widianto, A.A.S. Gunawan, Y. Heryadi, W. Budiharto, Evaluation of machine learning on smart home data for prediction of electrical energy consumption, in *2023 International Conference on Computer Science, Information Technology and Engineering (ICCoSITE)*, Jakarta, Indonesia (2023), pp. 434–439
25. E. Khaoula, B. Amine, B. Mostafa, Evaluation and comparison of energy consumption prediction models case study: smart home, in *The 3rd International Conference on Artificial Intelligence and Computer Vision (AICV2023)*, ed. by A.E. Hassanien et al. March 5–7, 2023. AICV 2023. Lecture Notes on Data Engineering and Communications Technologies 164 (Springer, Cham, 2023)
26. Y. Liang, P.K. Saha, Energy consumption forecasting based on long short-term memory neural network with realistic smart meter data, in *2022 IEEE Symposium Series on Computational Intelligence (SSCI)*, Singapore, Singapore (2022), pp. 1374–1379
27. D. Durand, J. Aguilar, M.D. R-Moreno, An analysis of the energy consumption forecasting problem in smart buildings using LSTM. Sustainability **14**(20), 13358 (2022)
28. M. Lichman, UCI machine learning repository (2013). https://archive.ics.uci.edu/ml/datasets/individual+household+electric+power+consumption
29. P. Carmona, F. Climent, A. Momparler, Predicting failure in the U.S. banking sector, An extreme gradient-boosting approach. International Review of Economics and Finance. (2018).

# IoT Device Fingerprinting for Anomaly Detection

**Justice Owusu Agyemang, Sylvester Junior Ampomah, Awumee Gabriel Selorm, Richard Somda Vinkpedomeh, and Patience Berko**

## 1 Introduction

A typical IoT device consists of a microcontroller, sensors, a communication module, clock oscillators, ADCs, GPIOs, memory, and a power source [1]. Many IoT devices are deployed in untrusted locations, making them vulnerable to counterfeiting and attacks. Therefore, it is essential to uniquely identify the microcontroller, sensors, and communication module.

In anomaly detection, network traffic is viewed as a collection of bidirectional flows, each identified by source and destination IP addresses, protocol number, and upper-layer identifiers [2]. Flows between transport layer endpoints are identified by source and destination IP addresses, transport protocol, and ports. Tools like Wireshark can capture these flows [3].

Existing fingerprinting techniques for sensor and RF communication module identification face challenges such as packet sniffing and the need for additional hardware. Devices are typically identified using IP and MAC addresses, but these can be spoofed to gain unauthorized access. Techniques like IP and MAC spoofing involve forging network identities to bypass access controls. For example, Gu et al. [3] explored the BLE protocol stack to extract unique network-flow characteristics for device authentication using internal clock skew, which is difficult to forge. However, this method is time-consuming and impractical for scenarios requiring rapid device authentication [1, 3].

J. O. Agyemang (✉) · S. Junior Ampomah · A. G. Selorm · R. S. Vinkpedomeh · P. Berko
Faculty of Electrical/Computer Engineering, Kwame Nkrumah University of Science and Technology, Kumasi, Ghana
e-mail: jay@knust.edu.gh; sjampomah@st.knust.edu.gh; sgawumee@st.knust.edu.gh; rvsomda@st.knust.edu.gh; pberko@st.knust.edu.gh

 
I. Woungang, S. K. Dhurandher (eds.), *The 7th International Conference on Wireless, Intelligent and Distributed Environment for Communication*,
Lecture Notes on Data Engineering and Communications Technologies 237,
https://doi.org/10.1007/978-3-031-80817-3_6

This study aims to develop a novel device fingerprinting technique using packet inter-arrival time (IAT) and received power level attributes from both physical and network layers. Combining these features, we aim to create a robust method for identifying IoT devices, enhancing security by making it harder for unauthorized entities to impersonate devices. IAT is distinct for each device due to hardware and software variations, making it a reliable fingerprinting feature [2]. Physical layer attributes further enhance this method's uniqueness and resistance to spoofing [2].

Our approach involves capturing packet data over wireless channels, with plans to explore wired channels. The project phases include setting up a test environment, data collection, and analysis, developing a deep learning model for device fingerprinting, and evaluating model performance. The goal is to ensure data integrity and secure network access for IoT devices by accurately identifying them and preventing unauthorized access.

The structure of this paper is as follows: Sect. 2 reviews related works and outlines the objectives. Section 3 details the proposed methods. Section 4 presents the results of the evaluation of these methods. Finally, Sect. 5 concludes the paper and offers recommendations for future work.

## 2 Related Works

The rapidly evolving landscape of the Internet of Things (IoT) has significantly heightened the need for robust and innovative security mechanisms to protect the integrity and privacy of interconnected devices. This literature review explores various contemporary approaches to IoT device fingerprinting and identification, with a particular focus on leveraging deep learning and other advanced techniques. As IoT devices become increasingly prevalent, they also become prime targets for cyberattacks, making the identification and authentication of these devices critical for maintaining secure communication channels.

Aneja et al. [4] developed a method for identifying IoT devices by analyzing packet inter-arrival times (IATs). They created unique signatures for each device without relying on network identifiers like IP addresses or physical identifiers like MAC addresses. This approach is particularly useful in wireless networks where communication is vulnerable to identity spoofing. However, the absence of standard datasets presents a significant limitation for benchmarking their method's effectiveness.

Vaidya et al. [5] focused on designing noninvasive, robust physical unclonable functions (PUFs) as device-specific identifiers. By leveraging inherent variations in component manufacturing, they created PUFs using off-the-shelf components. The resulting IoT-ID combines features from multiple PUFs, serving as a unique identifier and a primitive for generating secret keys. Despite the method's robustness, its dependency on specific hardware components limits its generalizability to a broad range of IoT devices.

Aversano et al. [6] analyzed packet header features and inter-arrival times to detect various attack types using machine learning models. Their approach, however, was constrained to specific packet types and did not generalize well to diverse attack scenarios. Li et al. [7] employed deep learning techniques for RF fingerprinting by converting RF signal samples into images for training convolutional neural networks (CNNs). This method demonstrated high accuracy in identifying RF signals and differentiating between genuine and spoofed devices, making it a robust solution for securing IoT networks. Bassey et al. [8] used deep learning to extract features from RF signals for device identification. Their approach involved clustering algorithms that effectively grouped devices based on their RF characteristics, although it did not directly address spoofing attacks. Gu et al. [3] explored the BLE protocol stack to extract unique network-flow characteristics from the link layer and ATT/GATT service layer for device authentication. This hardware-based fingerprinting approach relies on utilizing internal clock skew, which is difficult to forge but time-intensive, restricting practical application in scenarios requiring rapid fingerprinting.

Pan et al. [9] proposed a deep learning model that outperformed traditional supervised machine learning models in IoT device identification. By leveraging extensive network activity traces, the model demonstrated superior accuracy in identifying devices based on their unique network behaviors. Jafari et al. [10] focused on capturing RF data across varying signal-to-noise ratio (SNR) levels for wireless device identification. Using deep learning techniques, they achieved high identification accuracy even under challenging conditions. However, their approach was limited to the physical layer, which may not be sufficient for comprehensive security in complex IoT networks. In their 2024 study, Khan and Alkhathami focus on anomaly detection within IoT-based healthcare systems using advanced machine learning methods. They utilize the CIC IoT dataset, designed for IoT-specific attacks, to assess several models including random forest, adaptive boosting, and deep neural networks. Their findings highlight random forest's exceptional performance with an impressive accuracy rate in anomaly detection, surpassing other models tested [11]. To manage the computational demands of high-dimensional data, the authors employ feature reduction techniques, crucial for real-time healthcare applications. Despite promising outcomes, the study advocates for further exploration into real-time anomaly detection and handling diverse datasets to fortify IoT security in dynamic healthcare settings. Yue et al. [12] conducted a comprehensive survey on the application of deep learning in IoT security. They explored various deep learning techniques for enhancing IoT security, including device fingerprinting, anomaly detection, and intrusion prevention. Their evaluation demonstrated the significant potential of deep learning in addressing security and privacy challenges in IoT environments. Chen et al. [13] categorized various anomaly detection methodologies, focusing on TinyML, a subfield of ML for low-power devices, and its application in real-time anomaly detection using CNNs and autoencoders. They noted challenges such as implementing complex algorithms on constrained IoT hardware and the lack of real-world datasets for evaluation. Khan et al. [11] focused on anomaly detection in IoT-based healthcare systems using the CIC IoT dataset, evaluating models like random forest, adaptive boosting, and deep neural networks. They found random forest to be the most effective, with a 99.55% accuracy rate, and emphasized the need for further research into real-time anomaly detection and handling diverse datasets to enhance IoT security in healthcare.

# 3 Methodology

In this section, we provide an in-depth account of the research methods and procedures employed to address our study's objectives. We detail the systematic steps taken and the methodologies utilized in the pursuit of answering our research questions.

## *3.1 Data Collection*

To collect the data needed for this project, we set up a test bench with four Raspberry Pis, from two different vendors as illustrated in Fig. 1. We then configure a personal computer running Wireshark—The packet sniffer—as an access point (AP). The access point broadcast the beacon frames(sperixlab). The Raspberry Pis then connect through a wireless link to the AP. We configure Wireshark to capture channel information, specifically focusing on 802.11b radio information. Afterward, we enter the ping command to send ping packets to the raspberry pies to capture the network characteristics.

From the channel information, we extract the received power levels as contained in the 802.11b radio information by first converting the pcaps to a JSON file and using Python to extract the needed information as shown in Figs. 2 and 3. For the

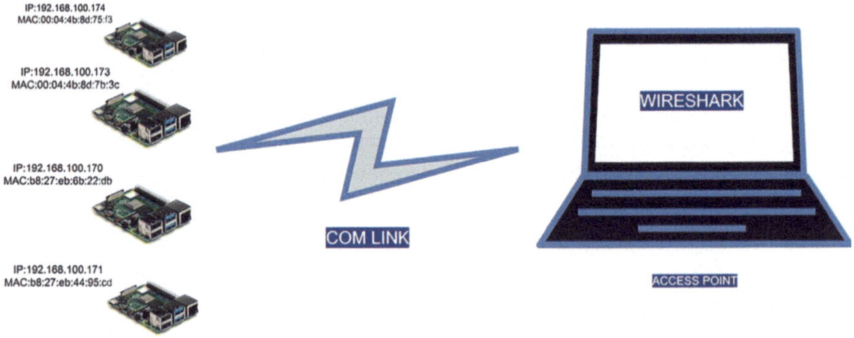


**Fig. 1** Bench test setup

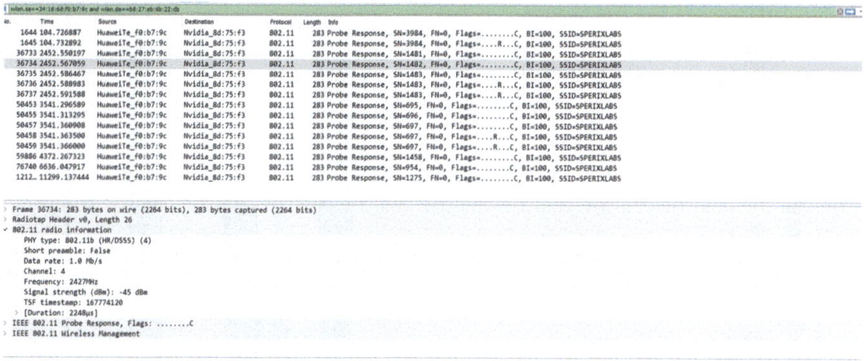

**Fig. 2** Wireshark interface-channel characteristics

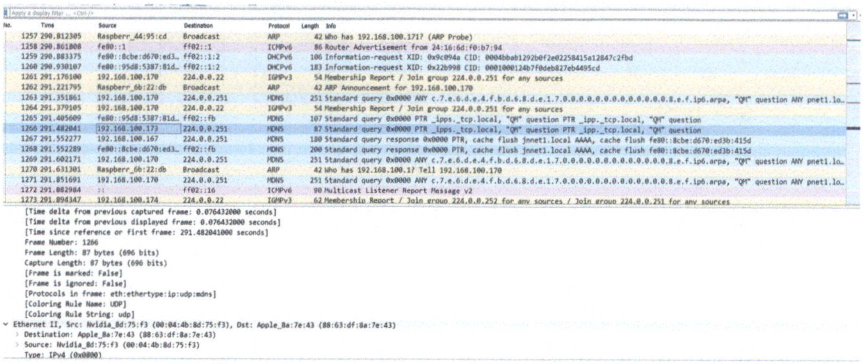

**Fig. 3** Wireshark interface network

network characteristics, we extract the packet inter-arrival time, which is the time difference from previous captured packets using the same approach as described above.

The dataset chosen for this study consists of both network and physical layer characteristics, specifically packet inter-arrival time (IAT) and received power level characteristics. This dual-layer approach was selected to provide a comprehensive and robust fingerprinting method that captures the unique behavior of IoT devices. The IAT is sensitive to hardware and software variations, making it a reliable feature for distinguishing between devices, while the physical layer attributes further enhance the uniqueness and resilience of the fingerprint against spoofing attempts. This combination aligns with our goal of improving the accuracy and reliability of device identification in IoT environments.

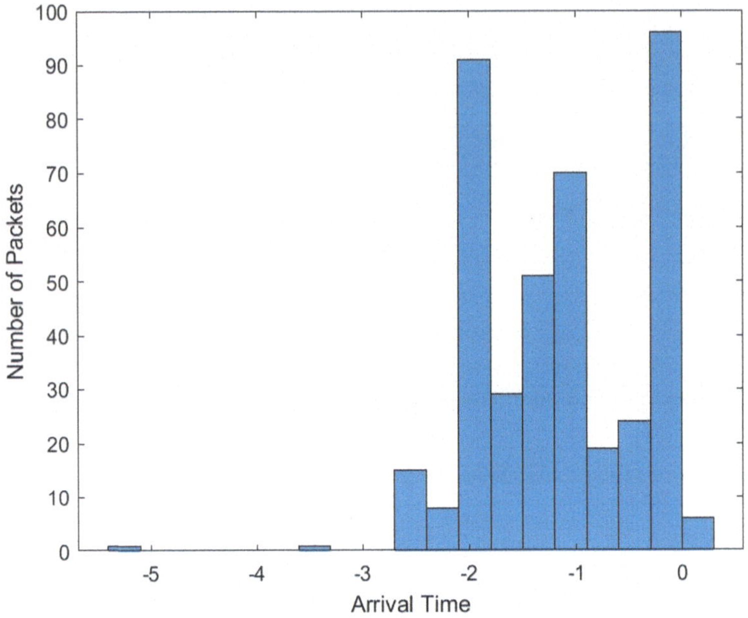


**Fig. 4** Packet inter-arrival time

## 3.2 Data Preprocessing

The packet inter-arrival time is first visualized in a histogram in Fig. 4, and due to the very small-time differences, we apply the log10 function to linearize it. The histogram is converted into a ".PNG" file to serve as an input for our CNN models.

The received power levels that characterize the channel information as faced by the device are converted to a scatter diagram with the number of packets on the vertical axis and the power levels—in dBm—on the horizontal axis. This is also converted into a ".PNG" file.

## 3.3 Model

With the objective of the paper being to use a CNN to recognize a device based on the IAT and received power level graphs, the graphs of the IAT and received power levels were plotted (Fig. 5). These plotted graphs were then used by the CNN for image classification to determine if the algorithm could distinguish between the devices solely based on those two plots without considering any other information. Since we are taking a few examples of IAT plots, we augmented the data by a

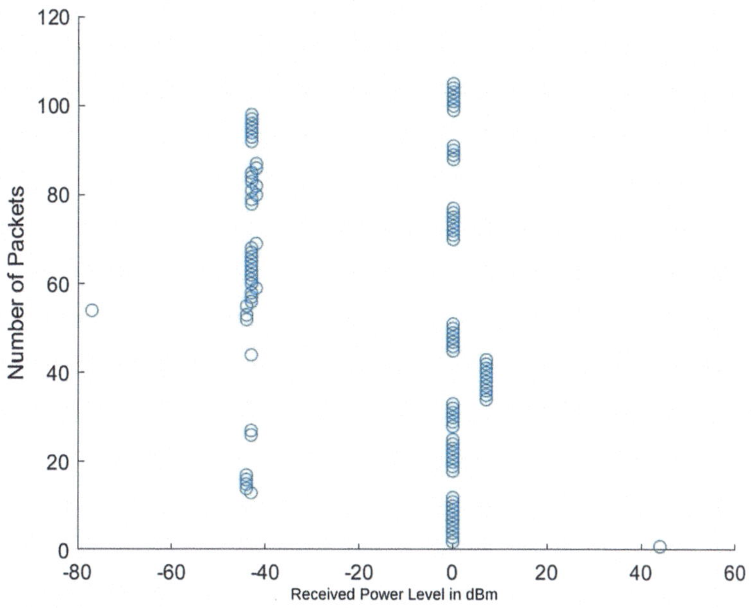


**Fig. 5** Power levels in dBm

number of random transformations so that the model does not see the same picture twice. This also prevents overfitting and generalizing the model.

The flowchart in Fig. 6 describes the whole fingerprinting process. It begins with extracting features in the form of IAT (inter-arrival time) graphs and signal-level graphs. These graphs, which include IP addresses as labels for known devices, are fed into a CNN (convolutional neural network) model. The CNN model processes these inputs to generate a categorized output. For known devices, the labeled graphs help in accurate classification. Unlabeled IAT and signal-level graphs are also processed by a trained CNN model, producing an output that may need further analysis. If the model's categorized output aligns with known behavior patterns, the device is classified as known. If it does not align, it triggers a decision-making process to determine if the behavior is recognizable. If recognized, the device is classified as known; otherwise, it is flagged as an anomaly, indicating an unknown or suspicious device. This system ensures continuous monitoring and classification of devices based on their network behavior.

## *3.4 Convolutional Neural Network*

A typical CNN architecture for image classification can be seen in Fig. 7. First, an input layer was declared to process the input images. The input shape parameter for

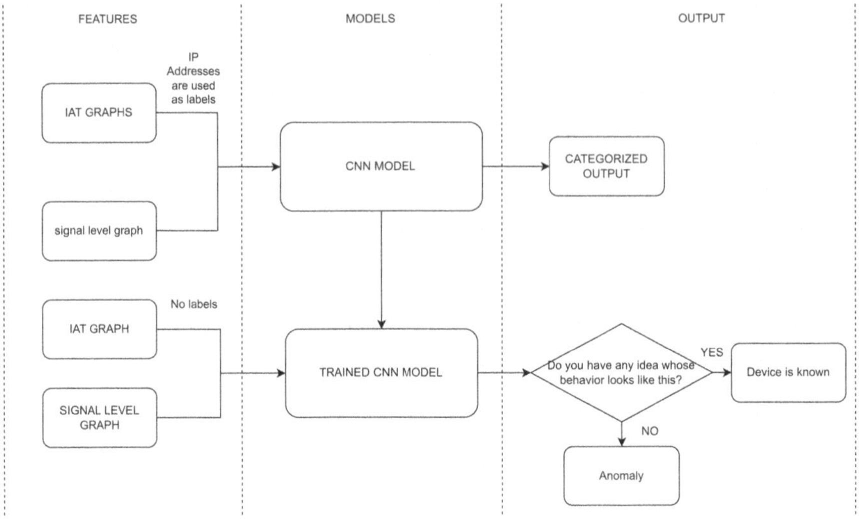


**Fig. 6** The fingerprinting flowchart

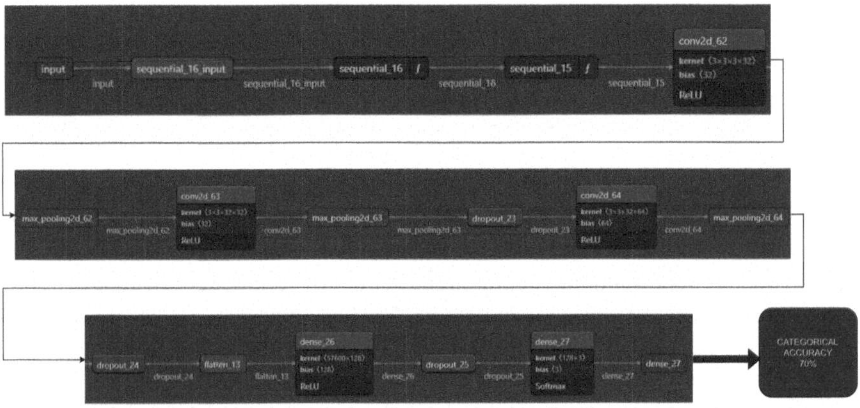


**Fig. 7** A CNN used for image classification

this layer was set to (3, 256, 256), corresponding to the depth, width, and height of the images, which were resized to 256×256 pixels. This shape specification is necessary only for the input layer. A set of 32 convolutional filters with a kernel size of 3×3 was employed, utilizing a rectified linear unit (ReLU) as the activation function. Subsequently, a 2D max-pooling layer with a pooling size of 2×2 was added, reducing the number of parameters by sliding a 2×2 window and selecting the maximum value from the four values in each window.

Next, two additional convolutional layers with 32 and 64 filters, respectively, were incorporated, each with a 3×3 kernel size and ReLU activation, followed by max-pooling layers with a 2×2 stride. The batch size was set to 32, indicating that training images would be processed in batches of 32, which helps manage memory usage and speeds up training by updating weights after each batch propagation. The number of epochs was set to 50, where one epoch consists of a complete forward and backward pass through all the training images.

### 3.5 Resnet50

The next model considered in this work is ResNet50 with pre-trained weights. ResNet50 was chosen for training due to the limited size of our dataset. Since ResNet50 has some of the neural weights already set, it speeds up the learning process. This approach is known as transfer learning [14].

### 3.6 Hyperparameter Tuning

To optimize the model's performance, we employed GridSearch as our hyperparameter tuning technique. This method systematically explored various combinations of hyperparameters to identify the most effective configuration for our specific task. The selected hyperparameters included the ReLU activation function for the intermediate layers, which helps prevent the vanishing gradient problem and enhances model training. For the final layer, Softmax was chosen as the activation function to handle the multi-class classification task. We used the stochastic gradient descent (SGD) optimizer with a learning rate of 0.001, balancing the speed of convergence with the need to avoid overshooting during training. A batch size of 100 was set, allowing for efficient memory management and faster training while processing batches of data. The model was trained over 50 epochs, which was sufficient for convergence while preventing overfitting.

### 3.7 Evaluation

The performance evaluation of the models in this study involved a comprehensive assessment using various metrics to determine their effectiveness in distinguishing between IoT devices. The evaluation process was based on fundamental concepts such as true positives (TP), true negatives (TN), false positives (FP), and false negatives (FN) classifications, as defined within the context of our dataset.

- **TN:** Cases where the model correctly classifies a known device's behavior as known.
- **TP:** Cases where the model correctly identifies an unknown device's behavior as anomalous.

- **FP:** Cases where the model incorrectly classifies a known device's behavior as anomalous.
- **FN:** Cases where the model incorrectly identifies an unknown device's behavior as known.
- **Total Number of Samples:** The total number of device behavior samples in the dataset.

#### 3.7.1 Accuracy

Accuracy is a fundamental metric used to evaluate the overall correctness of the model's predictions. It measures the proportion of correctly classified instances, including both true positives (**TP**, correctly identified IoT devices) and true negatives (**TN**, correctly identified other types of IoT devices), out of the total number of samples in the dataset. Accuracy provides an intuitive understanding of the model's correctness, especially in scenarios with balanced classes. The accuracy of an IoT device fingerprinting model can be calculated using the formula in Eq. (1):

$$\text{Accuracy} = \frac{\text{TP} + \text{TN}}{\text{Total Number of Samples}} \times 100\% \tag{1}$$

To holistically analyze the effectiveness of our proposed model, we conducted an extensive comparative assessment against prominent models documented in prior research studies. The results of this evaluation are presented in Table 1, which outlines the performance metrics of diverse models across various research endeavors.

## 4 Results and Discussion

### *4.1 Training and Testing*

The pcaps and the received power levels of each device were split into 0.8 and 0.2 for training and validation, respectively, and to reduce biases, the data was shuffled, with the shuffle size set to 50. Tables 1 and 2 show the results and the performance of each of the model.

**Table 1** Data partitioning

| Device IP address | Training data | Validation data | Total |
|---|---|---|---|
| 192.168.100.170 | 478 | 120 | 598 |
| 192.168.100.171 | 413 | 104 | 517 |
| 192.168.100.173 | 472 | 119 | 591 |
| 192.168.100.174 | 481 | 121 | 602 |

**Table 2** Model performance

| Model | Performance (categorical accuracy) |
|---|---|
| Convolutional neural network | 83% |
| ResNet-50 | 85.2% |

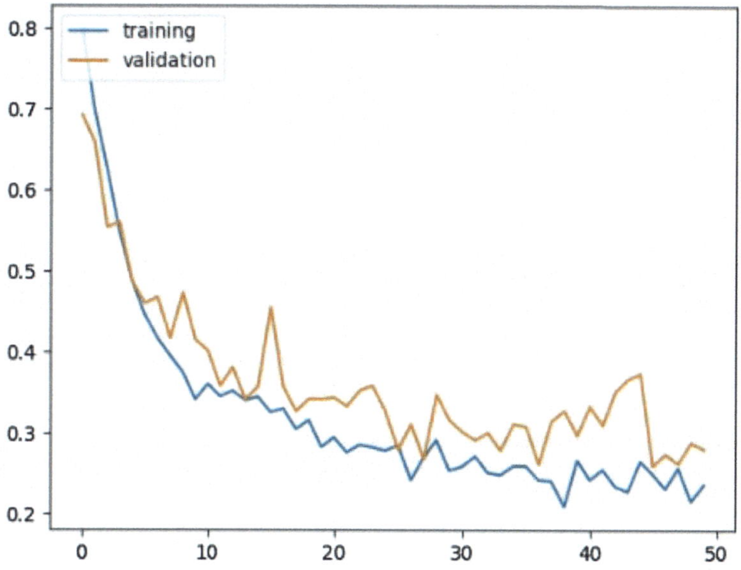


**Fig. 8** Model losses with increasing number of epoch

Figures 8 and 9 show the model accuracy and loss with an increasing number of epochs. Figure 9 illustrates the progress in accuracy as the number of epochs increases. We can observe that the model improves slightly after each epoch. Figure 8 shows how the loss in the model is reduced as the epochs increase. Loss, also known as mean squared error, is minimized after each epoch, which is evident in the graph below.

The results indicate that both the CNN and ResNet-50 models can effectively differentiate between IoT devices based on their network and physical layer characteristics. As seen in Table 2, ResNet-50 outperformed the CNN model slightly, achieving a categorical accuracy of 85.2% compared to the CNN's 83%.

Table 3 shows that our model has the lowest accuracy among the three compared models. Despite this, our method demonstrates promise due to its integration of both physical and network layer characteristics. This dual-layer approach leverages features from both layers to create a more robust model. However, its effectiveness relies on having sufficient data. With a larger and more diverse dataset, our model could potentially identify more intricate patterns and significantly enhance its accuracy.

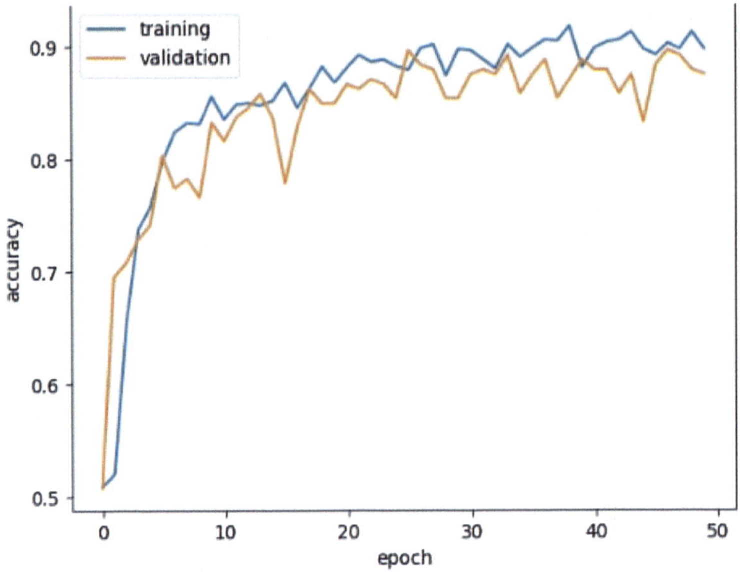


**Fig. 9** Model accuracy with increasing number of epoch

**Table 3** Comparison with related works

| Paper | Approach | Result |
|---|---|---|
| Our model | ResNet-50 | 85.2% |
| Jafari et al. [10] | Used RF data and deep learning for high-accuracy wireless device identification across different SNR levels | 90% |
| ANeja et al. [4] | Developed IoT device identification using packet inter-arrival times times (IATs), not relying on IP or MAC addresses | 86.7% |
| Richmond Osei et. al. [15] | Autoencoder with decision tree classifier | 97% |

# 5 Conclusion

In this paper, we focused on capturing packets sent from four Raspberry Pi devices to a laptop configured as an access point (AP) using Packet Capture (PCAP) library routines and a packet sniffer application. We developed a novel device fingerprinting (DFP) approach and conducted experiments on a dataset captured in real time. By applying both deep learning and machine learning techniques to this dataset, we achieved significant accuracy in device identification.

Our approach integrates both physical layer (received power level characteristics) and network characteristics (packet inter-arrival time), distinguishing it from previous research that often includes only one of these layers. By combining both physical and network characteristics, our method provides a more comprehensive and robust device fingerprinting solution. Specifically, our model achieved an accuracy of 85.2%. Although our models did not outperform some other existing models, our method demonstrates promise and robustness due to its integration of physical and network characteristics.

To further improve performance, we recommend collecting a larger and more diverse dataset. Utilizing Universal Software Radio Peripheral (USRP) to capture comprehensive channel characteristics could enhance the model's ability to identify IoT devices, particularly in open environments where wireless channels are prevalent. This expanded dataset would allow for the identification of more intricate patterns, potentially increasing the accuracy and robustness of our device fingerprinting approach.

## References


1. S. Aneja, M.S. Islam, IoT devices fingerprinting, in *2019 IEEE 5th World Forum on Internet of Things (WF-IoT)* (2019), pp. 661–666
2. S. Radhakrishnan, A. Ramasamy, A. Ramachandran, IoT device fingerprinting using packet inter-arrival time, in *2023 IEEE 5th International Conference on Cyber Security and Protection of Digital Services (Cyber Security)* (2023), pp. 1–6
3. Z. Gu, S. Luo, M. Zhang, Q. Yan, Y. Yang, Passive BLE fingerprinting for device authentication via inter-arrival time, in *GLOBECOM 2018—2018 IEEE Global Communications Conference* (2018), pp. 1–6
4. S. Aneja, N. Aneja, M.S. Islam, IoT device fingerprint using deep learning, in *2018 IEEE International Conference on Internet of Things and Intelligence System (IOTAIS)* (2018)
5. K. Vaidya, S. Ali, A. Shah, M.Z.A. Bhuiyan, W. Iqbal, Towards non-invasive robust physical unclonable function-based authentication for IoT devices. IEEE Internet of Things J. **8**(15) (2021)
6. L. Aversano, M.L. Bernardi, M. Cimitile, R. Pecori, A systematic review on deep learning approaches for IoT security. Comput. Sci. Rev. **40**, 100389 (2021)
7. B. Li, E. Cetin, Deep learning-based IoT device identification using RF signal images, in *2021 IEEE International Conference on Communications (ICC)* (2021), pp. 1–6
8. U. Bassey, J. Ovat, O. Adegoke, Intrusion detection in IoT networks using deep learning and RF signals. Internet Things J. 6(4) (2019)
9. J. Pan, Z. Wu, Z. Ma, H. Kim, IoT device fingerprint identification using deep learning. Comput. Secur. **104**, 102187 (2021)
10. H. Jafari, O. Omotere, D. Adesina, H.-H. Wu, L. Qian, IoT devices fingerprinting using deep learning, in *MILCOM 2018 - 2018 IEEE Military Communications Conference (MILCOM)* (2018)
11. M.M. Khan, M. Alkhathami, Anomaly detection in IoT-based healthcare: machine learning for enhanced security. Sci. Rep. **14**(1), 5872 (2024)
12. L. Yue, L. Deng, G. Li, Y. Liu, Deep learning for IoT security: a survey. IEEE Commun. Surv. Tutorials **23**(3) (2021)
13. L. Chen, X. Wang, Y. Jiang, K. Yang, X. Wang, J. Chen, Anomaly detection based on artificial Intelligence of Things: a systematic literature mapping. Patterns **5**(7), 100912 (2024)
14. S. Raut, Transfer learning with resnet50 in keras (2020). https://github.com/rautsrijana/Transfer_learning. Accessed 9 Apr 2024
15. R. Osei, H. Louafi, M. Mouhoub, Z. Zhu, Efficient IoT device fingerprinting approach using machine learning, in *Proceedings of the 19th International Conference on Security and Cryptography (SECRYPT 2022)* (SCITEPRESS - Science and Technology Publications, Lda, 2022), pp. 525–533

# Survey-Based Machine Learning Models for Early Detection of Diabetes

**Eric Brown and Wei Lu**

## 1 Introduction

Diabetes is a chronic disorder characterized by sustained high blood glucose levels. It substantially impacts organs and systems, including the heart, blood vessels, eyes, kidneys, and nerves, doubling the likelihood of vascular disorders in affected individuals [1]. Research also indicates a strong link between diabetes and certain cancers, such as liver cancer, as well as other nonvascular diseases [2, 3]. As a significant global health concern, diabetes was estimated by the International Diabetes Federation in 2019 to affect 463 million individuals worldwide, with projections suggesting a 51% increase by 2045 [4]. This global impact underscores the urgency of our research in developing effective early detection and prevention strategies.

Early diagnosis and treatment are vital for mitigating the substantial health risks associated with diabetes. Studies, like the ADDITION-Europe Simulation Model Study, have shown that early diagnosis can significantly reduce both absolute and relative risks of cardiovascular events and mortality [5]. A sensitivity analysis in the United States indicated that diagnosing diabetes two years earlier could reduce complication rates by 25%. The primary diagnostic methods for diabetes involve measuring blood glucose levels through invasive lab tests and home glucometer readings [6]. Despite their accuracy, these methods have several drawbacks, including high costs, time consumption, the need for specific technical skills, and the discomfort associated with blood sampling. Our proposed model, however, has the

E. Brown · W. Lu (✉)
Department of Computer Science, Keene State College, USNH, Keene, NH, USA

The University System of New Hampshire, Concord, NH, USA
e-mail: wlu@usnh.edu

 
I. Woungang, S. K. Dhurandher (eds.), *The 7th International Conference on Wireless, Intelligent and Distributed Environment for Communication*,
Lecture Notes on Data Engineering and Communications Technologies 237,
https://doi.org/10.1007/978-3-031-80817-3_7

potential to reduce significantly these health risks, offering a more optimistic future for diabetes diagnosis and prevention.

In contrast, artificial intelligence (AI) and statistical techniques like linear regression and advanced machine learning (ML) techniques offer a noninvasive, cost-effective, and efficient alternative for diabetes diagnosis. These techniques utilize real-world datasets, such as the Pima Diabetes dataset from the US National Institute of Diabetes and Digestive and Kidney Diseases, available at the University of California Irvine Machine Learning Repository, the Center for Disease Control (CDC) Behavioral Risk Factor Surveillance System (BRFSS) 2015 Diabetes Health Indicators Dataset, and the BIT Mesra Diabetes Dataset 2019. These datasets include various noninvasive symptomatic observations and measurements related to diabetes. However, it is crucial to note the absence of the gold standard diagnostic criteria, such as blood glucose levels, in these datasets, which poses challenges for data collection.

Although ML techniques leverage patterns in historical data to predict new cases, and integrating different machine learning techniques into ensemble models enhances predictive accuracy beyond individual models, the methods present challenges. There is considerable diversity in the machine learning methods employed across studies, complicating the identification of the most effective techniques. Additionally, a lack of transparency regarding the features used in these models hampers their interpretability, which is crucial for clinical application.

This chapter tackles these challenges and introduces a new survey-based machine learning approach for early diabetes detection. Our model employs the synthetic minority oversampling technique (SMOTE), alongside various machine learning methods, to enhance the performance of current algorithms. This innovation aids healthcare professionals in diagnosing diabetes and developing effective preventative measures. The remaining sections of the chapter are structured as follows: In Sect. 2, we summarize the existing literature on employing machine learning techniques to predict the survival rate of heart failure. Section 3 presents datasets from the CDC's BRFSS survey for 2015 and 2021, focusing on cleaning and merging data, including balancing and using techniques like SMOTE and synthetic minority oversampling technique with edited nearest neighbors (SMOTE-EEN). Sect. 4 focuses on our methodology, where we tested machine learning models on datasets of 2015 and 2021 using accuracy scores and confusion matrices across various configurations. After identifying the best model through Bayesian optimization, we grouped eight datasets for evaluation, ensuring a comprehensive framework for assessing model performance and robustness in diabetes classification. Section 5 discusses the experimental evaluation results, in which using the 2021 BRFSS dataset for training and the 2015 BRFSS for validation showed extreme gradient boosting (XGBoost) outperforming other models with lower false negatives (FNs) and false positives (FPs), indicating superior generalization. Also, implementing SMOTE increased accuracy significantly, reducing error classifications markedly in the SMOTE versus balanced datasets. The practical application of SMOTE, particularly evident in ensemble models like random forest, underscored its value in enhancing minority representation and overall model performance. Further optimizations using Bayesian search on combined datasets refined predictions, achieving an accuracy of 89.01%, demonstrating that extensive, representative data can improve model generalizability. Finally, Section 6 concludes the chapter with future works.

## 2 Related Works

Recent studies explore the effectiveness of machine learning (ML) algorithms in predicting and managing Type 2 diabetes (T2D) across various populations. Tigga and Garg highlighted the random forest classifier as the most accurate in assessing diabetes risk based on lifestyle and health data from 952 individuals [7]. Kaviyaadharshani et al. and Sai Krishna et al. both underscored the potential of ML in early diagnosis and personalized treatment plans, with techniques ranging from neural networks to an ensemble diabetic prediction algorithm (EDPA) showing promising results [8, 10]. Studies by Kour and Bharti and Liu et al. demonstrated the importance of early detection, employing algorithms like the extra tree classifier and XGBoost, which showcased high accuracy in specific datasets [9, 13]. Meanwhile, Jiang et al. focused on optimizing feature selection for risk prediction, identifying obesity as a significant factor [12]. Collectively, these studies emphasize the capability of ML to transform diabetes care through effective risk prediction, early detection, and tailored management strategies, thereby enhancing preventive healthcare and patient outcomes [11, 14, 15].

Tigga and Garg [7] focus on predicting Type 2 diabetes risk among individuals based on lifestyle, health, and family background through various machine learning algorithms. The study utilized data from 952 individuals collected via questionnaires alongside the Pima Indian Diabetes Database for comparative analysis. The study highlights the effectiveness of machine learning in accurately predicting diabetes risks, which is essential for early diagnosis and preventive healthcare. The random forest classifier emerged as the most accurate model among the tested algorithms, including logistic regression, k-nearest neighbors (kNN), support vector machine (SVM), naïve Bayes, and decision trees.

Kaviyaadharshani et al. [8] outline significant advancements in the early detection and management of diabetes through machine learning (ML) methodologies, where various ML models, such as random forests, support vector machines, and neural networks, are evaluated for their effectiveness in diagnosing diabetes. The research underscores the importance of innovative ML approaches to managing the complexity of diabetes diagnosis, which could significantly impact patient care by allowing for early intervention and personalized treatment plans.

Kour and Bharti [9] present a comprehensive study on utilizing machine learning techniques to improve the early detection of diabetes. This study emphasizes the importance of early detection, due to diabetes being an incurable disease that leads to severe health complications if not managed properly. The study analyzes data from 768 individuals, focusing on various attributes essential for predicting diabetes using machine learning algorithms. Several machine learning algorithms are employed to predict diabetes occurrence, including extra tree classifier, support vector machine (SVM), and random forest. The algorithms' performance is assessed based on accuracy and F1 scores. The extra tree classifier is highlighted for its superior accuracy (80%), compared to SVM. This indicates its potential effectiveness in diabetes detection applications.

Sai Krishna et al. [10] provide a comprehensive analysis of various machine learning (ML) algorithms applied to predict diabetes, focusing on enhancing early diagnosis accuracy. The study evaluates multiple ML algorithms, including decision trees (DT), support vector machines (SVM), and an ensemble diabetic prediction algorithm (EDPA), to identify the most effective approach for diabetes prediction. The results demonstrate the potential of machine learning in predicting diabetes effectively, with the EDPA model achieving the highest accuracy.

Bokhare and Vandan Raj [11] discuss using two prominent machine learning algorithms for diabetes prediction based on specific medical diagnostic measures. The study assesses the predictive capabilities of logistic regression and random forest algorithms using various performance metrics. The comparative analysis shows that logistic regression outperforms random forest in specific metrics, providing detailed insights into the strengths and weaknesses of each algorithm in diabetes prediction. Their study concludes by examining the effectiveness of machine learning algorithms in diagnosing diabetes early and emphasizing the importance of choosing the right algorithm for specific data types.

Jiang et al. [12] focus on optimizing the features used in diabetes risk prediction models through a comprehensive analysis of questionnaire data. The study addresses the increasing prevalence of diabetes and aims to refine diabetes risk prediction tools by optimizing the selection of characteristic variables in diabetes risk questionnaires. The focus is on enhancing the effectiveness of these questionnaires to improve the identification and management of diabetes risks among populations. Machine learning techniques were employed to assess the strength of correlations among diabetes risk factors using enumerated feature variables. The study successfully identified a ranking of characteristic variables that significantly impact diabetes risk, with obesity noted as the most impactful factor. Other factors like psychological elements, age, cholesterol levels, blood pressure, and socioeconomic status also showed varying degrees of correlation with diabetes risk. The results highlight the potential for optimizing questionnaire variables to improve the efficiency and accuracy of diabetes risk assessments.

Liu et al. [13] focus on developing effective machine learning (ML) models to predict Type 2 diabetes mellitus (T2DM) in the elderly Chinese population using a retrospective cohort study. The study utilized health screening data from 2018 to 2020 for adults older than 65 in Wuhan, China. The researchers developed prediction models based on logistic regression, decision trees, random forests, and extreme gradient boosting (XGBoost). The data included 127,031 records, with 8298 incident T2DM cases identified during the follow-up. Using LASSO regression, 21 predictive features were selected, addressing class imbalance with random under-sampling. Model performance metrics included Area Under the Curve (AUC), sensitivity, specificity, and accuracy. The XGBoost model showed the best performance with an AUC of 0.7805. Significant predictors identified were fasting plasma glucose, education, exercise, gender, and waist circumference. The study highlights the potential of the XGBoost model to screen high-risk individuals effectively.

Xie et al. [14] delve into creating effective predictive models to identify risk factors for Type 2 diabetes using machine learning, where the study utilizes data from the 2014 Behavioral Risk Factor Surveillance System (BRFSS), encompassing 138,146 participants, of which 20,467 had Type 2 diabetes. Machine learning algorithms employed include support vector machines, decision trees, logistic regression, random forest, neural networks, and Gaussian naive Bayes. All models demonstrated high predictive accuracy with AUC ranging from 0.7182 to 0.7949. The neural network model had the highest accuracy (82.4%), specificity (90.2%), and AUC (0.7949), while the decision tree model excelled in sensitivity (51.6%).

Md. Merajul Islam et al. [15] focus on using machine learning (ML) to predict Type 2 diabetes (T2D) and identify its risk factors. The study aims to identify risk factors for T2D using multiple logistic regression and to predict T2D using various ML techniques. The research utilized five ML techniques: logistic regression, naïve Bayes, J48, multilayer perceptron, and random forest. For 2009–2010, six risk factors were significant: age, education, marital status, systolic blood pressure (SBP), smoking, and body mass index (BMI). For 2011–2012, nine risk factors were identified: age, race, marital status, SBP, diastolic blood pressure (DBP), direct cholesterol, physical activity, smoking, and BMI. The random forest classifier showed the highest performance metrics, including accuracy, sensitivity, F-measure, and AUC.

## 3 BRFSS Dataset Preparation

The datasets used in this experiment were obtained from the Behavioral Risk Factor Surveillance System (BRFSS) of the Center for Disease Control (CDC), an annual health survey that collects responses from approximately 400,000 adults across the United States. Our analysis focused on data of the years 2015 and 2021. The pre-extracted datasets include "diabetes_binary_health_indicators_BRFSS2015.csv" and "diabetes_binary_health_indicators_BRFSS2021.csv" and consist of 23 features and did not undergo preprocessing or data balancing, thus maintaining a more significant number of records for each label [16]. The datasets feature a target variable, "diabetes_binary," used for binary classification, where "0" signifies no diabetes and "1" indicates prediabetes or diabetes.

After acquiring the datasets, we initiated the preprocessing stage, which included eliminating any null values, infinite values, or duplicate entries. We then changed the target variable's name from "diabetes_binary" to "diabetes" and categorized the BMI feature into six distinct groups. Following these adjustments, we had two cleaned datasets corresponding to 2015 and 2021. These datasets were subsequently merged into a single master dataset to facilitate testing.

In the testing phase, we structured the data into two principal categories. The first category included a combined dataset incorporating data from the 2015 and 2021 surveys. The second category utilized the 2021 dataset for training purposes and the 2015 dataset for validation. This approach allowed us to examine the results, thus deepening our insights thoroughly. We further subdivided each of these categories into four distinct sub-datasets.

The first subcategory consisted of a balanced dataset. This setup ensured equal entries for each classification in the combined 2021 datasets. Specifically, the 2021 dataset was balanced to have 33,395 records for each label, bringing the total to 66,790 records. Similarly, the combined dataset was adjusted to contain 68,238 records per label, resulting in 136,476 records. This balancing allowed for a fair comparison between the classifications in subsequent analyses.

The second subcategory utilized SMOTE, a method designed to generate synthetic data for the minority class, which, in this case, consisted of individuals diagnosed with diabetes. SMOTE was implemented on both the 2021 dataset and the combined dataset. For the 2021 dataset, the minority class was augmented from 33,395 to 189,848 records for each class, effectively balancing the dataset. Similarly, when SMOTE was applied to the combined dataset, it balanced the classes to 374,516 records each, ensuring an equal representation of both classes for analysis.

The third subcategory employed the synthetic minority oversampling technique with edited nearest neighbors (SMOTE-EEN), an enhanced version of SMOTE that combines oversampling with cleaning algorithms. This approach was applied to both the 2021 and combined datasets. For the 2021 dataset, this resulted in 106,554 entries for individuals without diabetes and 179,823 entries for those with diabetes, creating a deliberate imbalance in test model robustness. Similarly, the combined dataset after applying SMOTE-EEN showed 209,172 records for individuals without diabetes and 347,566 records for those with diabetes, further enabling analysis of model performance under varied data distributions.

The fourth subcategory consisted of a balanced dataset achieved using SMOTE-EEN. This method was selected to determine if balancing the dataset could alleviate imbalances and enhance the model's ability to generalize to new data. Initially, there were 209,172 records for individuals without diabetes, constituting the minority class. To achieve balance, we reduced the number of records in the diabetes class by 138,394, resulting in an equal count of 209,172 records for each class. After preparing these sub-datasets, we moved forward with the testing phase to evaluate the effectiveness of our models under these conditions.

## 4 Methodology

We measured the machine learning models' performance for all tests using an accuracy score and a confusion matrix to assess classification effectiveness. We tested the combined 2021 datasets, with the 2015 dataset as the validation set. Each dataset was tested across four configurations: a balanced dataset, a SMOTE dataset,

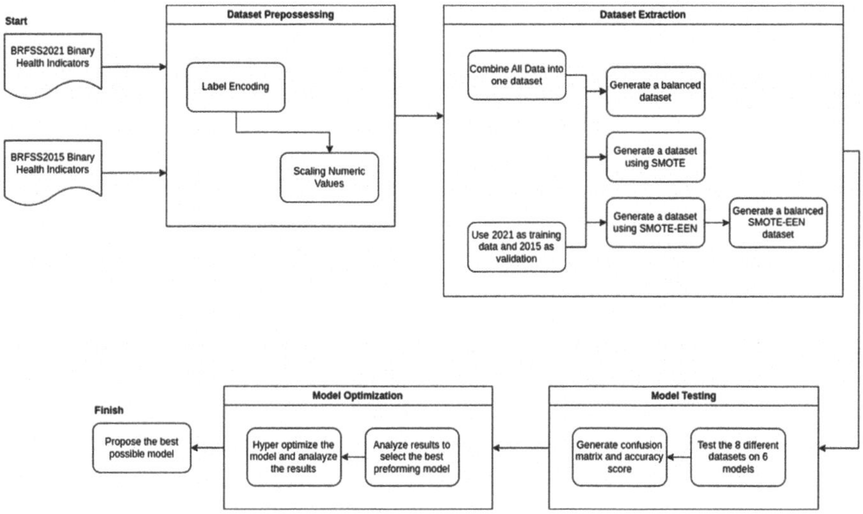


**Fig. 1** Flowchart of the methodology

a SMOTE-EEN dataset, and a balanced SMOTE-EEN dataset. This methodology provided a thorough analysis and enabled more precise interpretations of the results. Furthermore, to ensure the reliability of our results, we implemented fivefold cross-validation on the combined datasets. The 2015 validation set did not undergo cross-validation as it was utilized entirely for validation purposes. This structured approach allowed us to validate the models comprehensively and ensure robustness in our findings.

After having the best-performing category based on initial testing, we advanced to the hyper-optimization phase for the machine learning models. This phase employed Bayesian search optimization to fine-tune the top three models demonstrating performance. We conducted this optimization over 200 iterations, utilizing a fivefold cross-validation to ensure robust testing across expansive parameter spaces. The outcomes of this optimization were analyzed to select the best-performing model. Figure 1 illustrates a flowchart that outlines our comprehensive methodology, detailing each step from dataset preparation to the final optimization and selection process.

We collected eight datasets for our machine learning model evaluations during the dataset extraction phase. These datasets were organized into two groups for testing: one comprised all the combined data from both years, and the other utilized the 2021 dataset for training, with the 2015 dataset serving as the validation set.

## 5 Evaluation Results

The outcomes of employing the 2021 BRFSS dataset for training and the 2015 BRFSS dataset for validation are presented in Tables 1, 2, 3 and 4. These tables include accuracy scores and confusion matrices. Upon reviewing the accuracy scores, it becomes apparent that XGBoost consistently outperforms other models across all tests. Additionally, the confusion matrices reveal that XGBoost excels in generalizing new data, as indicated by its lower numbers of false negatives (FNs) and false positives (FPs).

Using the 2015 dataset as a validation set is essential when incorporating synthetic data generation methods such as SMOTE. This strategy ensures the validation dataset is synthetic data-free, maintaining consistency across all four testing categories. As a result, we observed a notable improvement in accuracy from the balanced dataset to the SMOTE dataset, with an increase from 67.83%

**Table 1** Results of 2015 balanced dataset

| Model type | Accuracy | Confusion matrix | |
|---|---|---|---|
| | | TP<br>FN | FP<br>TN |
| LR | 69.81% | 132,595<br>7468 | 61,782<br>27,611 |
| DT | 62.62% | 120,800<br>12,202 | 73,577<br>22,895 |
| RF | 67.39% | 126,782<br>7236 | 67,595<br>27,861 |
| KNN | 71.54% | 142,988<br>13,908 | 51,389<br>21,189 |
| XGBoost | 67.83% | 127,258<br>6711 | 67,119<br>28,386 |
| SVM | 66.53% | 123,541<br>5968 | 70,836<br>29,129 |

**Table 2** Results of 2015 SMOTE dataset

| Model type | Accuracy | Confusion matrix | |
|---|---|---|---|
| | | TP<br>FN | FP<br>TN |
| LR | 69.80% | 133,849<br>8779 | 60,528<br>26,318 |
| DT | 76.59% | 163,812<br>23,154 | 30,565<br>11,943 |
| RF | 83.51% | 182,986<br>26,440 | 11,391<br>8657 |
| KNN | 67.18% | 132,495<br>13,438 | 61,882<br>21,659 |
| XGBoost | 84.83% | 186,591<br>27,031 | 7786<br>8066 |
| SVM | 66.85% | 125,258<br>6952 | 69,119<br>28,145 |

**Table 3** Results of 2015 SMOTE-EEN dataset

| Model type | Accuracy | Confusion matrix | |
|---|---|---|---|
| | | TP<br>FN | FP<br>TN |
| LR | 57.95% | 101,558<br>3672 | 92,819<br>31,425 |
| DT | 69.35% | 136,721<br>12,688 | 57,656<br>22,409 |
| RF | 72.76% | 141,689<br>9830 | 52,688<br>25,267 |
| KNN | 58.57% | 105,842<br>6532 | 88,535<br>28,565 |
| XGBoost | 73.68% | 143,666<br>9692 | 50,711<br>25,405 |
| SVM | 56.29% | 97,139<br>3064 | 97,238<br>32,033 |

**Table 4** Results of SMOTE-EEN balanced dataset

| Model type | Accuracy | Confusion matrix | |
|---|---|---|---|
| | | TP<br>FN | FP<br>TN |
| LR | 64.30% | 118,077<br>5631 | 76,300<br>29,466 |
| DT | 72.11% | 144,939<br>14,573 | 49,438<br>20,524 |
| RF | 76.15% | 151,965<br>12,325 | 42,412<br>22,772 |
| KNN | 60.10% | 109,731<br>6907 | 84,646<br>28,190 |
| XGBoost | 77.73% | 156,161<br>12,889 | 38,216<br>22,208 |
| SVM | 63.41% | 115,503<br>5080 | 78,874<br>30,017 |

to 84.83%. Additionally, the confusion matrix for the balanced dataset indicated 73,830 FN and FP classifications, while the SMOTE model reduced this number to 34,817 FN and FP classifications, a significant reduction of 39,013 records.

The confusion matrix from the balanced dataset revealed 74,831 FN and FP classifications for the random forest model. With the SMOTE model, this number decreased to 37,831—a precise reduction of 37,000 records. This significant change underscores the effectiveness of SMOTE in enhancing the representation of minority labels. Furthermore, it is evident that ensemble machine learning models, such as random forest, achieve better performance on unseen data when trained with SMOTE-enhanced datasets.

When evaluating the SMOTE-EEN results against the balanced dataset outcomes in Tables 1, 2 and 3, we noted both decreases and increases in accuracy. The accuracy of supervised learning models such as logistic regression (LR), k-nearest neighbors (KNN), and support vector machine (SVM) declined. This trend suggests

**Table 5** Results of balanced dataset

| Model type | Accuracy | Confusion matrix | |
|---|---|---|---|
| | | TP<br>FN | FP<br>TN |
| LR | 64.30% | 48,497<br>16,670 | 19,741<br>51,568 |
| DT | 63.56% | 43,953<br>25,450 | 24,285<br>42,788 |
| RF | 72.01% | 46,844<br>16,808 | 21,394<br>51,430 |
| KNN | 66.42% | 52,509<br>30,079 | 15,729<br>38,141 |
| XGBoost | 73.51% | 47,033<br>14,945 | 21,205<br>53,293 |
| SVM | 73.50% | 46,177<br>14,108 | 22,061<br>54,130 |

**Table 6** Results of SMOTE dataset

| Model type | Accuracy | Confusion matrix | |
|---|---|---|---|
| | | TP<br>FN | FP<br>TN |
| LR | 73.78% | 264,292<br>88,836 | 109,574<br>287,680 |
| DT | 84.32% | 320,420<br>63,355 | 54,096<br>311,161 |
| RF | 88.45% | 353,028<br>65,012 | 21,488<br>309,504 |
| KNN | 82.07% | 244,573<br>4383 | 129,943<br>370,133 |
| XGBoost | 88.56% | 359,511<br>70,668 | 15,005<br>303,848 |
| SVM | N/A | N/A | |

that these models struggle with generalizing new data, particularly when synthetic data are involved, highlighting their limitations in such contexts. On the other hand, ensemble methods like decision trees (DT), random forest (RF), and XGBoost exhibited noticeable improvements in accuracy on the validation set. These observations confirm that XGBoost was the most effective model, followed by RF and DT, underscoring the strength of ensemble methods in handling synthetic data scenarios.

Next, the combined dataset's results are presented in Tables 5, 6, 7 and 8, including the accuracy score and confusion matrix. Although fivefold cross-validation was employed, including synthetic data might impact the authenticity of the results. Notably, even though there was an increase in accuracy from Table 5 (AllData_Balanced) to Table 8 (AllData_SMOTE) for the logistic regression (LR) and k-nearest neighbors (KNN) models, the counts of false positives (FPs) and false negatives (FNs) remained high. This contrasts with the results in Tables 1 to 4, where

**Table 7** Results of SMOTE-EEN dataset

| Model type | Accuracy | Confusion matrix | |
|---|---|---|---|
| | | TP<br>FN | FP<br>TN |
| LR | 85.39% | 161,947<br>34,118 | 47,225<br>313,448 |
| DT | 89.53% | 185,259<br>34,365 | 23,913<br>313,202 |
| RF | 92.05% | 196,305<br>31,391 | 12,867<br>316,175 |
| KNN | 93.89% | 176,513<br>1335 | 32,659<br>346,231 |
| XGBoost | 87.26% | 195,671<br>57,433 | 13,501<br>290,133 |
| SVM | 85.73% | 161,193<br>31,448 | 47,979<br>316,118 |

**Table 8** Results of SMOTE-EEN balanced dataset

| Model type | Accuracy | Confusion matrix | |
|---|---|---|---|
| | | TP<br>FN | FP<br>TN |
| LR | 84.56% | 175,585<br>31,004 | 33,587<br>178,168 |
| DT | 91.31% | 190,415<br>17,616 | 18,757<br>191,556 |
| RF | 94.79% | 201,065<br>13,694 | 8107<br>195,478 |
| KNN | 92.23% | 180,800<br>4122 | 28,372<br>205,050 |
| XGBoost | 94.35% | 201,353<br>15,833 | 7819<br>193,339 |
| SVM | 85.10% | 176,668<br>29,814 | 32,504<br>179,358 |

the synthetic data used in the validation set led to increased records, potentially skewing the results.

After thorough analysis, we concluded that employing the combined dataset with SMOTE is the most effective strategy. Our tests showed that using the 2021 dataset with the 2015 validation set and applying SMOTE significantly enhances model accuracy. Specifically, the XGBoost model achieved an accuracy of 84.84% with SMOTE-enhanced data. To further refine the model, we conducted 200 iterations of Bayesian search optimization on the 2021 and 2015 datasets, which resulted in an improved accuracy of 89.01%. This decision was guided by the understanding that a model trained on more representative data is better equipped to generalize to new data, leading us to utilize all available combined data for the best predictive performance.

The model optimization phase was conducted using all available data, leveraging the principle that models typically perform better with more data for classification.

**Table 9** Results of XGBoost optimization

| XGBoost optimization | |
|---|---|
| Parameters | Value |
| colsample_bytree | 0.40304 |
| gamma | 0.0 |
| learning_rate | 0.001 |
| max_depth | 118 |
| n_estimators | 97 |
| Accuracy score | 89.51% |

We identified discrepancies among Tables 1, 2, 3, 4, 5, 6, 7 and 8, where the results based on real data from Tables 1, 2, 3 and 4, contradicted the outcomes depicted in Tables 5, 6, 7 and 8. This discrepancy led us to choose SMOTE over SMOTE-EEN for synthetic data generation. Subsequently, we fine-tuned the model to enhance performance, employing a fivefold cross-validation during the optimization to aim for a more accurate evaluation, ensuring that synthetic data did not skew the results. The detailed parameter values used in this optimization process are displayed in Table 9. Through these efforts, we achieved an impressive accuracy score of 89.51%.

## 6 Conclusion and Future Work

In this chapter, we propose a survey-based model to facilitate the early diagnosis of diabetes, designed for seamless integration into a user-friendly interface. Patients would enter their names and complete a survey. To enhance accessibility, we propose the creation of a QR code that directly links to the online survey. It is crucial, however, to approach this method with stringent security measures to safeguard patient data during Internet transmission. Upon completion of the survey, the data will be analyzed using our machine learning model to produce a predictive assessment of the patient's diabetes risk.

In future work, we aim to broaden our research to encompass the detection of various health-related conditions, utilizing data from the BRFSS dataset, which includes information on numerous diseases. To enhance the effectiveness of preventative healthcare, we plan to develop a sophisticated model that employs clustering approaches [17, 18]. By clustering similar patient data and health conditions, we can more accurately and efficiently refine our risk assessment models for multiple common diseases. This advanced framework will allow patients to complete surveys either online or in a hospital setting, enabling the evaluation of their risk across different health conditions. Implementing clustering techniques, such as k-means or co-clustering, will improve the model's ability to identify patterns and correlations in large datasets, thereby providing more personalized and precise risk evaluations [19, 20]. Such a tool will empower patients with valuable health insights and facilitate early intervention and the development of tailored health management strategies.

**Acknowledgments** Research is supported by New Hampshire-INBRE through an Institutional Development Award (IDeA), P20GM103506, from the National Institute of General Medical Sciences of NIH.

## References

1. American Diabetes Association, 2. Classification and diagnosis of diabetes: Standards of medical Care in Diabetes-2020. Diabetes Care **43**(Suppl 1), S14–S31 (2020). https://doi.org/10.2337/dc20-S002
2. H.B. El-Serag, H. Hampel, F. Javadi, The association between diabetes and hepatocellular carcinoma: A systematic review of epidemiologic evidence. Clin. Gastroenterol. Hepatol. **4**(3), 369–380 (2006). https://doi.org/10.1016/j.cgh.2005.12.007
3. J.S. Kasper, E. Giovannucci, A meta-analysis of diabetes mellitus and the risk of prostate cancer. Cancer Epidemiol. Biomarkers Prev. **15**(11), 2056–2062 (2006). https://doi.org/10.1158/1055-9965.EPI-06-0410
4. International Diabetes Federation, *Diabetes* (International Diabetes Federation, Brussels, 2019)
5. W.H. Herman, W. Ye, S.J. Griffin, R.K. Simmons, M.J. Davies, K. Khunti, G.E. Rutten, A. Sandbaek, T. Lauritzen, K. Borch-Johnsen, M.B. Brown, N.J. Wareham, Early detection and treatment of type 2 diabetes reduce cardiovascular morbidity and mortality: A simulation of the results of the Anglo-Danish-Dutch study of intensive treatment in people with screen-detected diabetes in primary care (ADDITION-Europe). Diabetes Care. **38**(8), 1449–1455 (2015). https://doi.org/10.2337/dc14-2459. Epub 2015 May 18. PMID: 25986661; PMCID: PMC4512138
6. T.K. Mathew, P. Tadi, *Blood Glucose Monitoring* (Stat-Pearls Publishing, Treasure Island, 2021)
7. N.P. Tigga, S. Garg, Prediction of type 2 diabetes using machine learning classification methods. Procedia. Comput. Sci. **167**, 706–716, ISSN 1877-0509 (2020). https://doi.org/10.1016/j.procs.2020.03.336
8. D. Kaviyaadharshani, M. Nivedhidha, R. Jeyarohini, J. Lece Elizabeth Rani, M.P. Ramkumar, G.S.R. Emil Selvan, Diagnosing diabetes using machine learning-based predictive models. Procedia. Comput. Sci. **233**, 288–294 (2024). https://doi.org/10.1016/j.procs.2024.03.218
9. S.P. Kour, V. Bharti, Diabetes diagnosis using machine learning algorithms: Methods and techniques, in *International conference on SMART Generation Computing, Communication and Networking (SMART GENCON)*, vol. 2022, (Bangalore, 2022), pp. 1–5. https://doi.org/10.1109/SMARTGENCON56628.2022.10083626
10. T.V.S. Krishna, S.K. Pathuri, K.S. Sandeep, M.K. Padhi, I. Aswani, D. Haritha, An efficient machine learning classification model for diabetes prediction, in *IEEE 12th International Conference on Communication Systems and Network Technologies (CSNT)*, vol. 2023, (Bhopal, 2023), pp. 139–146. https://doi.org/10.1109/CSNT57126.2023.10134615
11. A. Bokhare, N.V. Raj, Diabetes prediction using logistic regression and random forest algorithm: A comparative study, in *International Conference for Advancement in Technology (ICONAT)*, vol. 2023, (Goa, 2023), pp. 1–5
12. L. Jiang, Z. Yang, G. Liu, Z. Xia, G. Yang, H. Gong, J. Wang, L. Wang, A feature optimization study based on a diabetes risk questionnaire. Front. Public Health **23**(12), 1328353 (2024). https://doi.org/10.3389/fpubh.2024.1328353. PMID: 38463161; PMCID: PMC10920272
13. Q. Liu, M. Zhang, Y. He, L. Zhang, J. Zou, Y. Yan, Y. Guo, Predicting the risk of incident type 2 diabetes mellitus in Chinese elderly using machine learning techniques. J. Pers. Med. **12**(6), 905 (2022). https://doi.org/10.3390/jpm12060905. PMID: 35743691; PMCID: PMC9224915
14. Z. Xie, O. Nikolayeva, J. Luo, D. Li, Building risk prediction models for type 2 diabetes using machine learning techniques. Prev. Chronic Dis. **19**(16), E130 (2019). https://doi.org/10.5888/pcd16.190109. PMID: 31538566; PMCID: PMC6795062

15. M.M. Islam, M.J. Rahman, M. Menhazul Abedin, B. Ahammed, M. Ali, N.A.M.F. Ahmed, M. Maniruzzaman, Identification of the risk factors of type 2 diabetes and its prediction using machine learning techniques. Health Sys. **12**(2), 243–254 (2022). https://doi.org/10.1080/20476965.2022.2141141
16. Annual Survey Data, https://www.cdc.gov/brfss/annual_data/annual_data.htm
17. W. Lu, L. Xue, A heuristic-based co-clustering algorithm for the internet traffic classification, in *28th International Conference on Advanced Information Networking and Applications Workshops*, (Victoria, 2014), pp. 49–54
18. W. Lu, H. Tong, I. Traore, E-Means: An Evolutionary Clustering Algorithm, in *Advances in Computation and Intelligence. ISICA 2008*, Lecture Notes in Computer Science, ed. by L. Kang, Z. Cai, X. Yan, Y. Liu, vol. 5370, (Springer, Berlin/Heidelberg, 2008)
19. W. Lu, I. Traore, Determining the optimal number of clusters using a new evolutionary algorithm, in *17th IEEE International Conference on Tools with Artificial Intelligence (ICTAI'05)*, (Hong Kong, 2005), pp. 2–713. https://doi.org/10.1109/ICTAI.2005.57
20. W. Lu, I. Traore, A new evolutionary algorithm for determining the optimal number of clusters, in *International Conference on Computational Intelligence for Modelling, Control and Automation and International Conference on Intelligent Agents, Web Technologies and Internet Commerce (CIMCA-IAWTIC'06)*, (Vienna, 2005), pp. 648–653

# Optimizing Wearable Sensors with Multifeature Approximate Computing

**Nishanth Chennagouni and Qiaoyan Yu**

## 1 Introduction

Driven by the demand from the aging of society and the growing chronic diseases such as cardiovascular disease, diabetes, respiratory disorders, and other underlying conditions, wearable sensor systems receive increasing attentions. The Health 4.0 concepts advocate leveraging Internet of Everything (IoE), wireless communication, and cyber-physical system infrastructures to advance the virtualization of individuals, medical devices, and customized online health monitoring [1]. In the new health-monitoring framework shown in Fig. 1, sensors and sensor computing systems play a vital role. The review [2] highlights that 42% of research literature with the keyword *sensor* published in IEEE, Science Direct, and PubMed databases in the period of January 2014–March 2020 is related to health monitoring. Among those works, wearable sensors take a 14% share. The work [3] shows that the market of mobile health quadrupled in 2 years, from 6.95 billion US dollars (USD) in 2017 to 25.17 billion USD. It is predicted that there will be 175 zettabytes of digital data by 2025, among which a significant portion is contributed by health-monitoring tools [4]. There is a pressing need to consider efficient ways to store, transfer, and analyze health-related big data, especially in the era of Health 4.0. Approximate computing has demonstrated its potential in the computer world and image processing, but it is still in its youth in other domains. We envision that approximate computing will be a promising solution to address the pressing need for health-monitoring applications. There is limited work available to investigate what proper approximation mechanisms are suitable for sensor systems employed

N. Chennagouni (✉) · Q. Yu
Department of Electrical and Computer Engineering, University of New Hampshire, Durham, UK
e-mail: NishanthGoud.Chennagouni@unh.edu



I. Woungang, S. K. Dhurandher (eds.), *The 7th International Conference on Wireless, Intelligent and Distributed Environment for Communication*,
Lecture Notes on Data Engineering and Communications Technologies 237,
https://doi.org/10.1007/978-3-031-80817-3_8

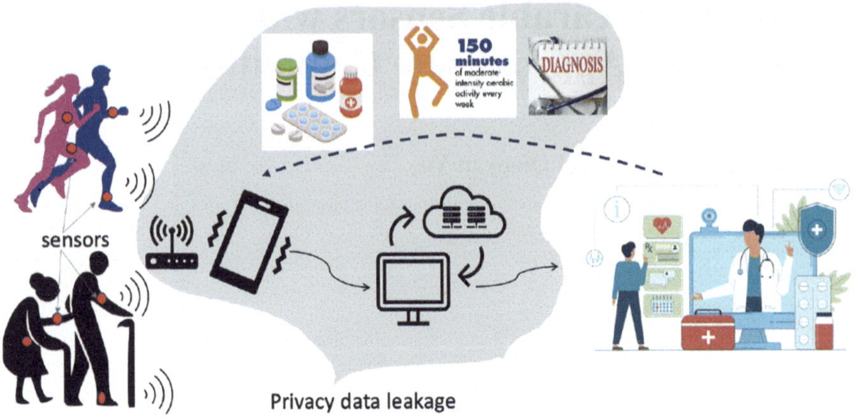


**Fig. 1** Application of wearable sensors in health-monitoring systems

in wearable health-monitoring systems. As privacy is always a concern in health applications, it is important to understand how approximation will affect privacy. To fill the gaps above, we propose a multifeature-driven approximation (MFDApx), which reduces the size of sensing data and the cost of data storage and transmission without sacrificing the information (i.e., *features*) useful for applications. More specifically, this work makes the following contributions:

- We extend our prior single feature-driven approximation (FDApx) [5] to a multifeature-driven approximation (MFDApx) in health-monitoring sensor systems. We simultaneously consider multiple features that will guarantee the accuracy of activity recognition.
- We apply the approximation concept to near-sensor computing systems to reduce the size of sensing data for communication and also preserve user data privacy.
- This is an early work that explores the feasibility of using an approximation to address the privacy leak issue in health-monitoring sensor networks.

The rest of this work is organized as below. The proposed method is presented in Sect. 2. The method assessment based on a published sensor dataset is provided in Sect. 3. This work is concluded in Sect. 4.

## 2 Proposed Multifeature-Driven Approximation (MFDApx) Method

### 2.1 *Overview of Proposed Method*

Existing wearable sensors measure biometrics (e.g., heart beats, blood pressure, and motion data) and transfer the raw data to mobile devices for preliminary processing in user apps such as Fitbit, VeryFit, SilverCrest Active, and VF Watch.

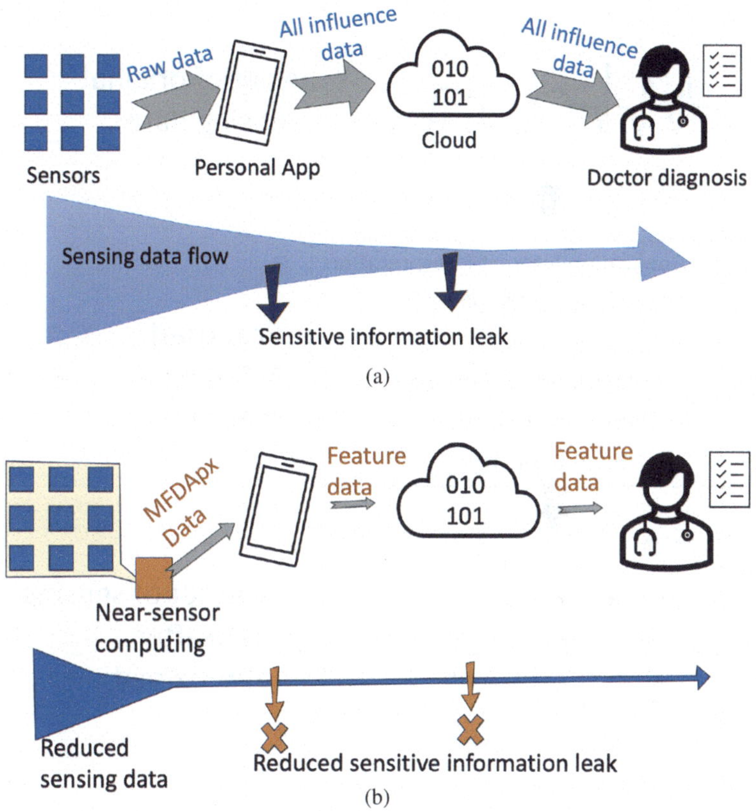


**Fig. 2** Sensing data flow in (**a**) non-approximate system and (**b**) proposed feature-driven approximation in a near-sensor computing system

More advanced apps can further assist professionals in making diagnoses, as shown in Fig. 2a. Although personal mobile devices will selectively send sensor data to the cloud, a large volume of real-time monitoring data floods to the later process stages, demanding massive storage space and challenging the protection mechanism against sensitive information leakage. In this work, we propose a multifeature-driven approximation (MFDApx) mechanism for near-sensor computing systems to significantly reduce the amount of sensor data transmission from sensors to its expected personal mobile app, as shown in Fig. 2b. To reduce sensitive information leakage in the middle of data transfer, our method only transmits data that carry features interested by the health-monitoring application, rather than all raw sensor data.

Our prior FDApx [5] analyzes the feature interested in the application to perform partial filtering on the massive raw sensor data. In this work, we extend that idea by identifying *multiple* features carried by a set of heterogeneous sensors in a wearable health-monitoring system and develop an advanced approximate computing

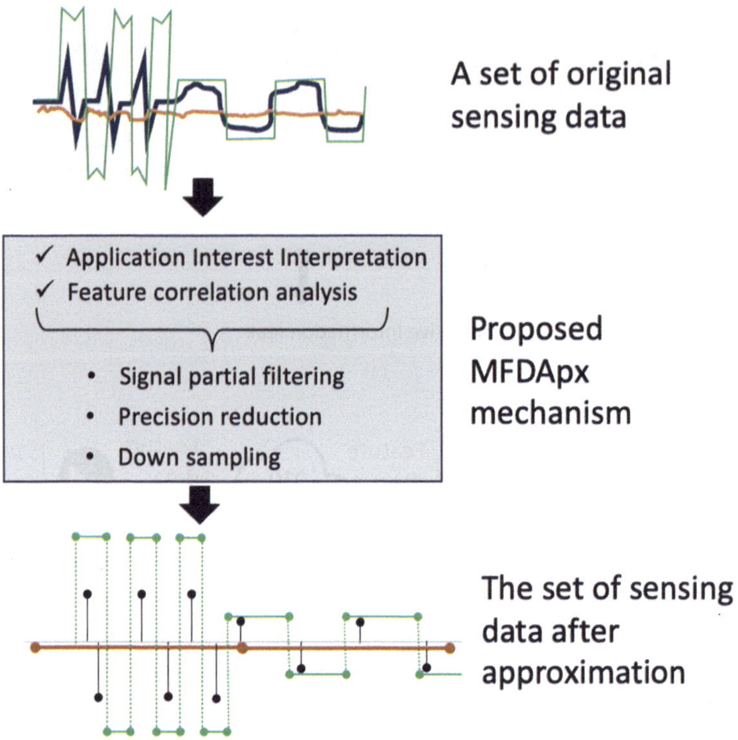


**Fig. 3** Concept of MFDApx

method. The concept of MFDApx is depicted in Fig. 3. First, we investigate the features interested in high-level applications and interpret the obtained features as the critical patterns that should be observed from the raw sensing data. At this step, we drive the thresholds for signal partial filtering. Next, we study the relative relationship among the values measured by heterogeneous sensors within the same sensing system. After that, we determine to what extent the precision can be reduced and the maximum time interval for sampling that will not lead to a noticeable loss of feature-critical information. As can be seen from Fig. 3, our MFDApx retains the critical data points and simplifies the switching patterns. Therefore, our approximation method can reduce the size of sensing data without losing information relevant to applications.

## *2.2 Proposed Feature-Driven Approximation Strategies*

Wearable sensors that monitor users' heartbeat, blood pressure, motion direction, and acceleration can provide strong clues for professionals to diagnose the patient's health conditions. In this work, we adopt a public database *MotionSense* [6] to

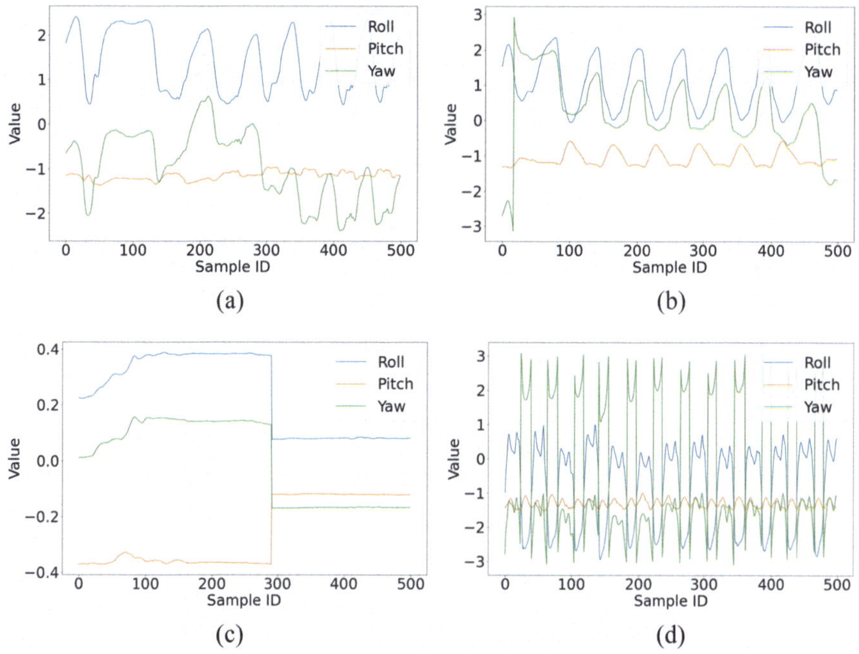


**Fig. 4** Attitude data for four activities: (**a**) sitting, (**b**) standing, (**c**) walking, and (**d**) jogging

explain our method. That dataset includes time-series data generated by accelerometer and gyroscope sensors (attitude, gravity, user acceleration, and rotation rate). A gyroscope is a device used for measuring or maintaining orientation and angular velocity. An accelerometer is a device used to measure changes in speed or direction. A few hundred samples of gyroscope and accelerometer sensing data are shown in Figs. 4 and 5, respectively. As can be seen, the data patterns for sitting, standing, walking, and jogging are significantly different in terms of switching patterns, absolute magnitude, relative magnitude within a sensing dataset, and data transition trends. The essence of our proposed feature-driven approximation is to use the features of interest to guide the minimization of the loss function $\Xi$ between the original sensor data $S$ and the approximated sensor data $\widetilde{S}$. More specifically, we have two approximation strategies to implement the approximation and minimize the information loss:

$$\Xi = loss\ function\left(S, \widetilde{S}_{MFDApx}\right) \underset{Feature}{\longmapsto} 0 \tag{1}$$

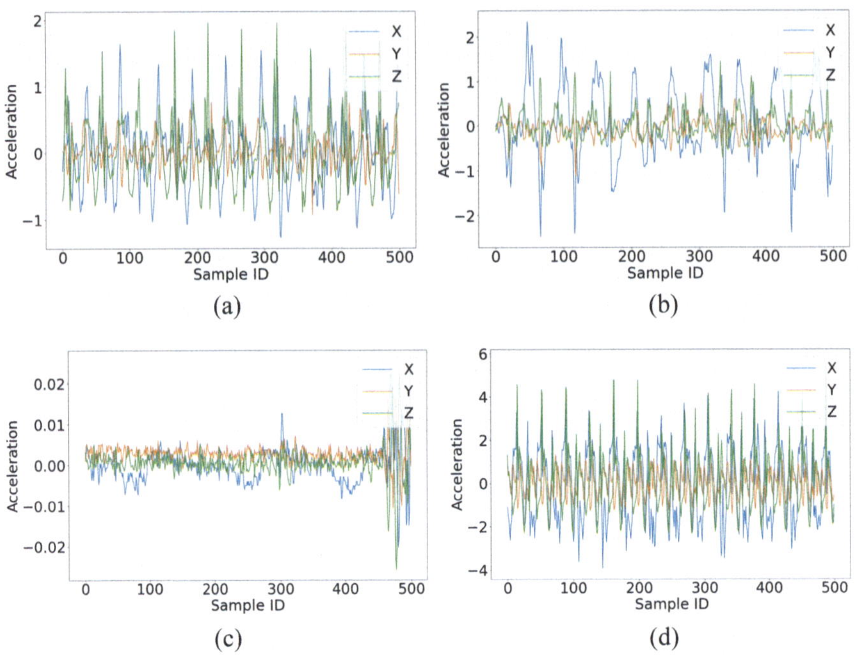


**Fig. 5** Acceleration on X, Y, and Z directions for activities: (**a**) sitting, (**b**) standing, (**c**) walking, and (**d**) jogging

### 2.2.1 Approximation Strategy #1

Jogging has the highest switching frequency among all four activities. If we only need to differentiate jogging from other kinds of activities, it is unnecessary to record the attitude data with full precision. This is because the switching frequency is sufficient to identify the jogging activity. *This observation inspires our approximation strategy 1: applying a customized ceiling/floor function* defined in Eq. (2):

$$\widetilde{S}_{MFDApx1} = \begin{cases} \lceil s \rceil & S > \theta_{p,c} \\ 0 & \theta_{n,c} < S < \theta_{p,c} \\ \lfloor s \rfloor & S < \theta_{n,c} \end{cases} \tag{2}$$

where $\theta_{p,c}$ and $\theta_{n,c}$ are the activity thresholds that indicate whether to apply a ceiling function $\lceil\cdot\rceil$ or a floor function $\lfloor\cdot\rfloor$, respectively. The thresholds $\theta$ are derived from the databases and categorized into multiple subsets:

$$\theta \in \left\{\Theta \mid \theta_{jogging}, \theta_{walking}, \theta_{sitting}, \theta_{standing}\right\} \tag{3}$$

**Fig. 6** Attitude data for female jogging activities (**a**) before and (**b**) after proposed approximation strategy #1

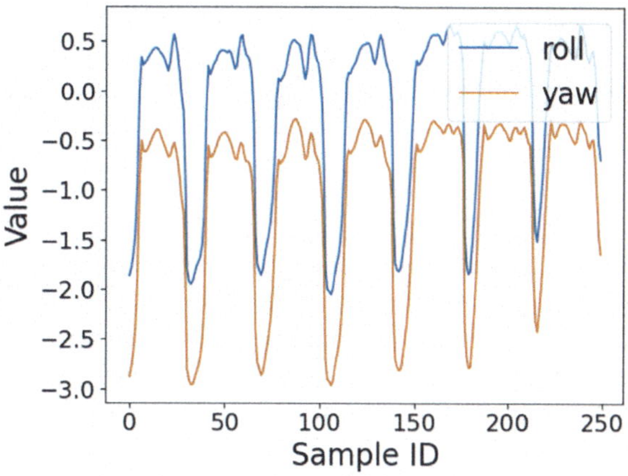


(a)

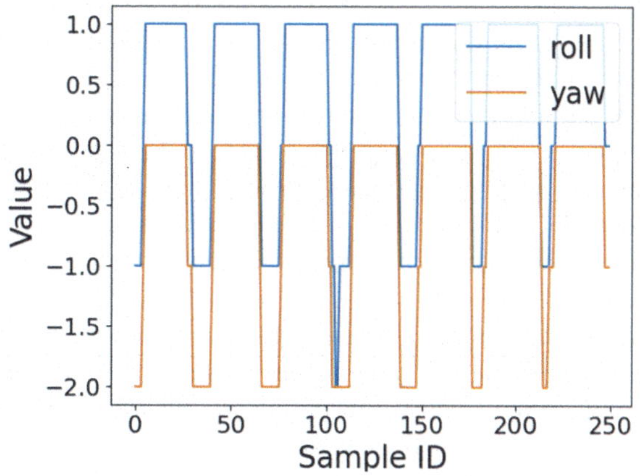


(b)

For instance, the peak value of roll and yaw is not necessary to differentiate a specific activity; thus, we can ignore the local vibration around positive and negative peaks, while we minimize the loss function. Using the MotionSense database [6], we illustrate the approximated roll and yaw attitude in Fig. 6. The experimental results provided in Sect. 4 show that the minimized loss function guarantees a negligible loss in the accuracy of activity recognition.

### 2.2.2 Approximation Strategy #2

For simplicity, the existing health-monitoring sensors use the same sensing frequency throughout the entire sensing period. However, this is not desirable because different sensing data have different transition speeds. This means that we should not apply the same sensing resolution to all the sensors. *This observation inspires our approximation strategy 2: applying selectively sampling interval to different sensors* defined in Eq. (4):

$$\widetilde{S}_{MFDApx2} = S[1 : \delta_s : End] \tag{4}$$

where $\delta_s \in \{\delta_{jogging}, \delta_{walking}, \delta_{sitting}, \delta_{standing}\}$. Those time intervals vary with the features interested in health-monitoring applications. For example, to determine the duration of different activities, the monitoring app only needs to recognize the activity, rather than knowing the quantitative magnitude of each activity's measurement. Thus, we can eliminate the repetitive sampling data. Once the activity pattern is confirmed, we can apply the sampling-based approximation to the sensing data.

As shown in Figs. 7 and 8, the attitude for walking transitions with a much slower speed than that for other activities and only jogging requires a fast sampling rate. Therefore, we do not need a high-speed transmission rate for walking. Furthermore, we can tailor the sampling rate for different activities. More specifically, we can employ different sampling rates to perform feature-driven approximation on the roll attitude data and x-direction acceleration data for both walking and jogging activities. As shown in Fig. 7a and b, the sampled data for walking is still well aligned with the original sensing data. We calculated the Pearson correlation coefficients between the original sensing data and its approximated version. As shown in Table 1, the correlation coefficients for the walking cases are 0.9875 and 0.9647 for $\delta_s = 5$ and $\delta_s = 20$, respectively. In contrast, as indicated in Figs. 8a and b, the sampling-based approximation will lead to more information loss in the sensing data of jogging than walking. If the sampling frequency is reduced to 1/5 of the original sensing speed, the offset on attitude and acceleration leads the correlation coefficient to drop to 0.8758. As we further reduce the sampling frequency to 1/20, the sampling approximation results in the Pearson correlation

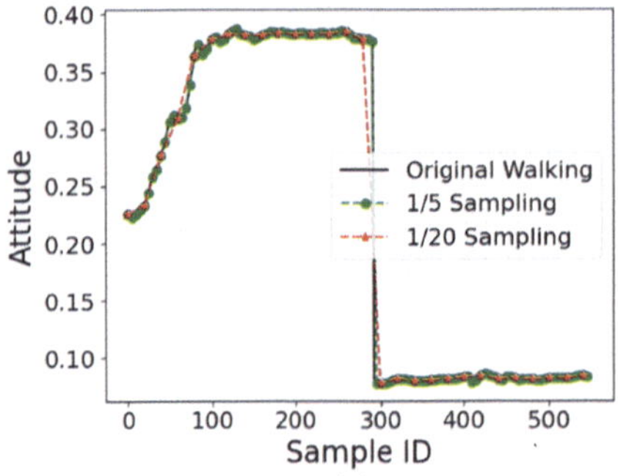

(a)

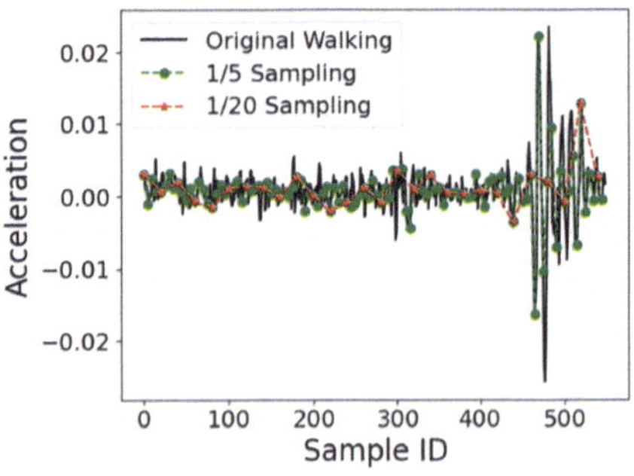

(b)

**Fig. 7** Comparison of (**a**) roll attitude value and (**b**) x-direction acceleration speed for the walking activity sensing with/without proposed approximation strategy #2

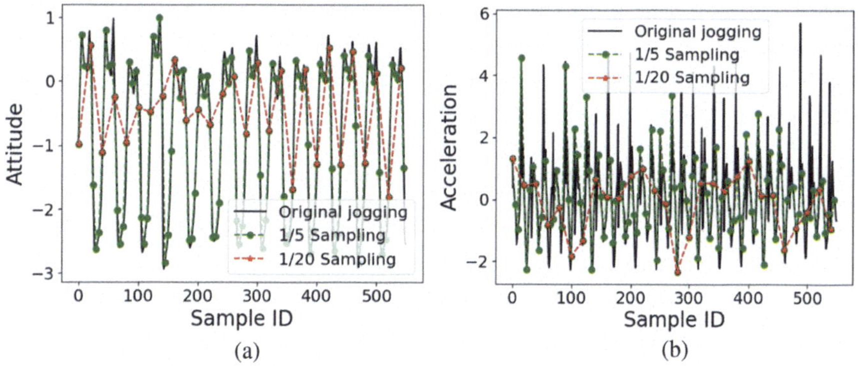


**Fig. 8** Comparison of (**a**) roll attitude value and (**b**) x-direction acceleration speed for the jogging activity sensing with/without proposed approximation strategy #2

**Table 1** Correlation coefficient reduction due to approximation

| Sampling rate/activity | Walking | Jogging |
|---|---|---|
| $\delta_s = 5$, 1/5 sampling | 0.9875 | 0.8758 |
| $\delta_s = 10$, 1/10 sampling | 0.9666 | 0.5036 |
| $\delta_s = 20$, 1/20 sampling | 0.9647 | 0.5253 |

coefficient decrease to 0.5253. As a result, we should adopt $\delta_{walking} = 20$ and $\delta_{jogging} = 5$ to balance the accuracy of activity recognition and the approximation-induced data reduction.

## 3 Experimental Results

### *3.1 Experimental Setup*

The MotionSense database [6] includes time-series data generated by accelerometer and gyroscope sensors (attitude, gravity, user acceleration, and rotation rate). The data was collected with an iPhone 6s kept in the participant's front pocket using SensingKit, which collects information from the Core Motion framework on iOS devices. All data are collected in 50Hz sample rate. A total of 24 participants in a range of gender, age, weight, and height performed 6 activities in 15 trials in the same environment and conditions: downstairs, upstairs, walking, jogging, sitting, and standing. In this work, we focus on the following four activities: walking, jogging, sitting, and standing.

We trained a multitask convolutional neural network (MTCNN) to recognize activity and gender on a specific dataset. Besides the learning with the original dataset, we evaluate our approximation method with various configurations listed in Table 2. To measure privacy leaks, we used test data that contained information

**Table 2** Different feature-driven approximation configurations assessed in this work

| Approximation method | Description |
|---|---|
| MDFApx1a | Precision reduction on all input data |
| MDFApx1b | Precision reduction only on jogging data |
| MDFApx1c | Precision reduction on jogging and walking data |
| MDFApx2a | Sampling rate reduction on all input data ($\delta_{jogging}, \delta_{walking}, \delta_{sitting}, \delta_{standing}$) = 10 |
| MDFApx2b | Sampling rate reduction on jogging data ($\delta_{jogging} = 10$) |
| MDFApx2c | Sampling rate reduction on jogging and walking data ($\delta_{walking} = 20$ and $\delta_{jogging} = 10$) |
| MDFApx2d | Sampling rate reduction on jogging and walking data ($\delta_{walking} = 20$ and $\delta_{jogging} = 5$) |

from previously unseen users while assuming an adversary had access to the training dataset. We evaluated the ability of a pre-trained MTCNN on user raw data (baseline) to predict user identity when provided with approximated data.

Our MTCNN consists of four convolutional layers, including convolutions with kernel sizes of 5, 3, 5, and 3 and depths of 50, 50, 40, 40, and 20, respectively. Max-pooling layers with pool sizes of 2 and 3 aid in downsampling. Dropout layers with dropout probabilities of 0.2 and 0.4 are strategically placed to avoid overfitting. After the convolutional layers, dense layers with a hidden size of 400 further increase feature extraction and classification. The model uses categorical cross-entropy and binary cross-entropy loss functions for activity recognition and gender classification, respectively. The architecture outputs predictions for both tasks through separate output layers, employing softmax activation for activity recognition and sigmoid activation for gender classification. This meticulously designed architecture aims to efficiently extract relevant features from input data while simultaneously performing accurate multitask classification.

## 3.2 Impact of Different Approximation Strategies on Classification Accuracy

Accuracy and F1 score are used to evaluate the impact of proposed approximation strategies on the activity recognition. The trained MTCNN model can recognize all four activities with an accuracy and F1 score close to 96.5%. If we approximate all the sensing data by removing its fraction portion (MFDApx1a), both accuracy and F1 score start to decrease, as shown in Fig. 9a. However, if we selectively apply precision reduction based approximation to the sensing data (MFDApx1b and MFDApx1c), the accuracy and F1 score are improved back to the levels that the baseline achieves. This result indicates that the approximation strategy that *selectively* reduces the data precision is able to preserve the classification accuracy.

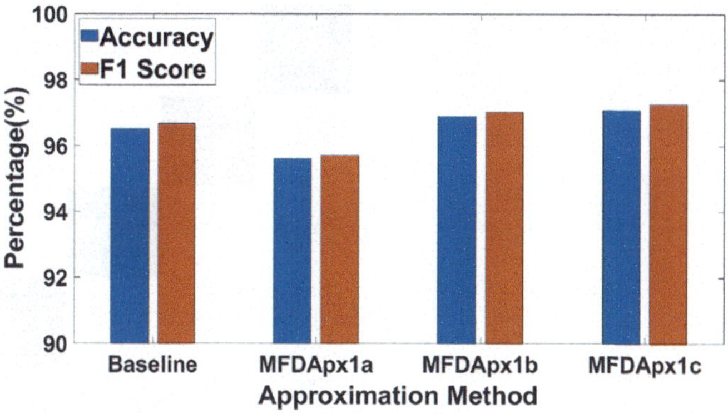


(a)

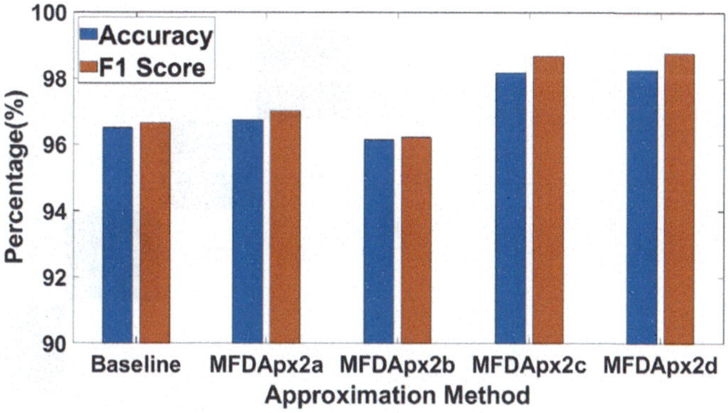


(b)

**Fig. 9** Accuracy of activity recognition (**a**) MFDApx1 and (**b**) MFDApx2

Figure 9b shows the impact of downsampling based approximation strategy on the accuracy and F1 score. If we blindly apply downsampling to sensor data (e.g., MFDApx2b), we could lose the accuracy of recognizing the correct activity. More interestingly, when the sensing data for both jogging and walking is pre-processed by MFDApx2c (from 50Hz to 10Hz and 10Hz, respectively) and MFDApx2d (from 50Hz to 5Hz and 20Hz, respectively), the accuracy and F1 score of the classifier increase to 98%. This experiment confirms that different sensors should be downsampled with a different rate, in order to obtain the same level of classification accuracy.

We further zoom in the approximation-induced change on the accuracy by studying confusion matrices shown in Fig. 10. As can be seen from Fig. 10a and b, the selective precision reduction (MFDApx1c) misclassifies more walking activity and sitting activity than the baseline by 2.65% and 3.97%. In contrast, the selective downsampling (MFDApx2c) eliminates all false positives and false negatives.

**Fig. 10** Confusion matrix: (**a**) baseline, (**b**) MFDApx1c, (**c**) MFDApx2c. wlk, walking; jog, jogging; sit, sitting; std, standing

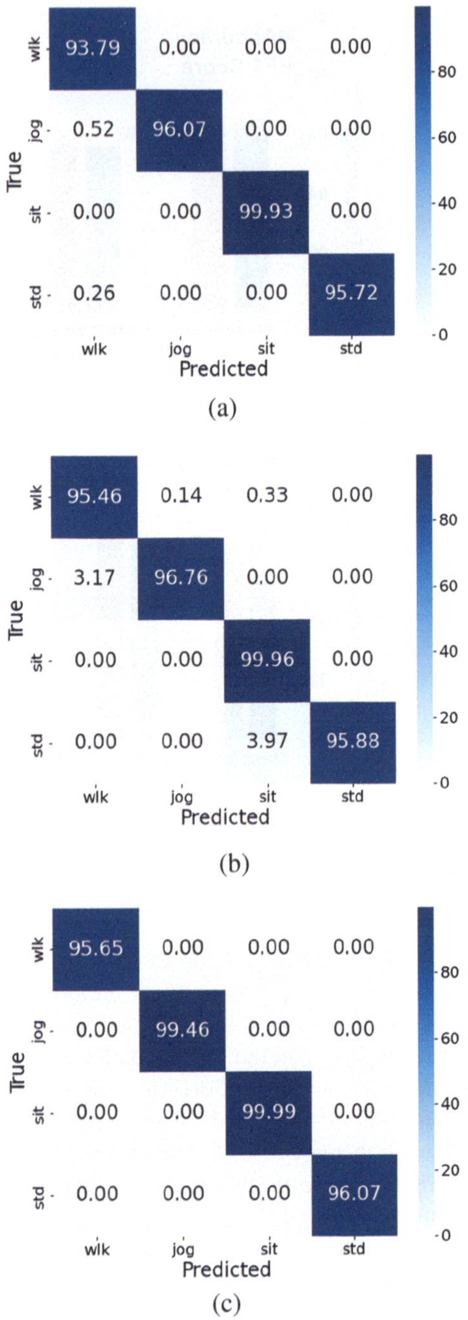

### 3.3 *Impact of Different Approximation Strategies on Privacy Leakage*

Approximate computing reduces the data precision or data density, but it may lead to more privacy leakage. In this subsection, we examine the privacy aspect by classifying the gender associated with the sensor readings in four activities. As shown in Fig. 11a, the precision-reduction-based approximation (MFDApx1) causes more privacy leakage. If the precision reduction is applied to all the sensor data, the leakage on gender increases by 4%. When we exploit the prior knowledge on the correlation between jogging activity and gender to selectively apply precision reduction approximation, we can reduce the privacy leakage down to 2%. This result means that some details in the sensor data (e.g., vibration in peak values) contribute to protect privacy.

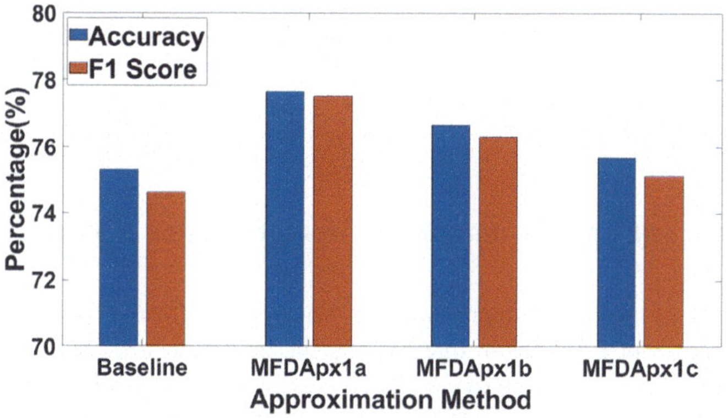


(a)

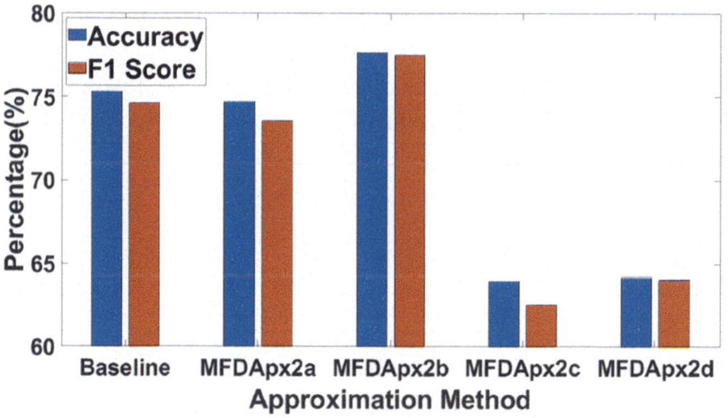


(b)

**Fig. 11** Impact of approximation on privacy leakage (gender recognition) (**a**) precision-reduction-based approximation (MFDApx1) and (**b**) downsampling-based approximation (MFDApx2)

**Table 3** Reduction on memory space achieved by proposed MFDApx

| Approximation method | Memory (MB) for each sensor |
|---|---|
| Baseline | 214.75 |
| MFDApx1a | 107.37 |
| MFDApx1b | 214.72 |
| MFDApx1c | 214.71 |
| MFDApx2a | 21.50 |
| MFDApx2b | 109.75 |
| MFDApx2c | 107.14 |
| MFDApx2d | 105.57 |

A similar trend is observed in Fig. 11b. Selective downsampling-based approximation can effectively prevent significant privacy leakage. The baseline, MFDApx2a, and MFDapx2b have an average accuracy and F1 score of 75%. In MFDApx2c and MFDApx2d, we employ different downsampling rates to the jogging and walking sensor data and thus reduce the accuracy and F1 score to 63%, which is 12% lower than the baseline. This experimental result indicates that selectively downsampling can reduce the size of sensor data and meanwhile preserve the privacy.

### *3.4 Impact of Approximation on Memory Space Saving*

The methods proposed for approximating sensor data effectively reduce its size. This study has four activities of interest, and the total data size for these activities amounts to 214.75 MB. Removing the fractional part from all input data sizes reduces the memory to 107.37 MB, as shown in Table 3. However, applying selective feature-driven fractional removal approximation does not reduce the data size compared to the baseline. On the other hand, sampling reduction approximation on all input data saves 89.98% memory size. Further selective downsampling memory reduces to 109.75 MB, 107.14 MB, and 105.57 MB for MDFApx2b, MDFApx2c, and MDFApx2d, respectively. This reduction is approximately 48.89%, 50.1%, and 50.75%, less than the baseline.

## 4 Conclusion

Wearable health-monitoring sensors generate significantly large volume of data, which challenges storage spaces in edge device and cloud. Approximate computing emerges as a promising solution to address the pressing need. This work proposes a multifeature-driven approximation method to reduce the size of sensor data. The multiple features interested in the high-level application guide the selection of approximation strategies, which leverage precision reduction and downsampling to condense sensor data without losing the accuracy of activity classification. The experimental results show that the proposed MFDApx can reduce the memory storage space by about 50% over the raw sensor data. The prior knowledge on the features of activity sensing value can preserve privacy leakage while retaining the classification accuracy.

**Acknowledgments** This project was partially supported by NSF Awards 2022279 and 2154606.

## References

1. J. Li, P. Carayon, IISE Trans. Healthc Syst. Eng. **11**(3), 171 (2021). https://doi.org/10.1080/24725579.2021.1884627
2. L.M.S.d. Nascimento, L.V. Bonfati, M.L.B. Freitas, J.J.A. Mendes Junior, H.V. Siqueira, S.L. Stevan, Sensors **20**(15) (2020). https://doi.org/10.3390/s20154063. https://www.mdpi.com/1424-8220/20/15/4063
3. mHealth Market mHealth Market Size, Share & Industry Analysis, By Category (By Apps Disease & Treatment Management, Wellness Management, By Wearable Body & Temperature Monitors, Glucose Monitors) By Services Type (Monitoring Services, Diagnostic Services, Treatment Services) By Service Provider (mHealth App Companies, Hospitals, Health Insurance) and Regional Forecast, 2019–2026 (2020). https://www.fortunebusinessinsights.com/industry-reports/mhealth-market-100266
4. P. Dhunnoo, Hackers, breaches and the value of health data: 2024 e-book update (2024). https://medicalfuturist.com/your-privacy-in-the-digital-health-era-the-medical-futurists-guide/
5. N. Chennagouni, M. Kajol, D. Chen, D. Xu, Q. Yu, in *Proceedings of the Great Lakes Symposium on VLSI 2024* (2024), pp. 712–717
6. M. Malekzadeh, R.G. Clegg, A. Cavallaro, H. Haddadi, in *Proceedings of the International Conference on Internet of Things Design and Implementation*, IoTDI '19 (ACM, New York, 2019), pp. 49–58. https://doi.org/10.1145/3302505.3310068

# Breast Cancer Histology Image Segmentation Using Connected Component Analysis

**Vincent Majanga and Ernest Mnkandla**

## 1 Introduction

Breast cancer remains the prime cancer-associated disease burden for women. It affects 1 in 20 worldwide and as many as 1 in 8 in high-income countries [1]. Breast cancer is a cancer that originates from breast cells. According to [2], cancer starts to develop in ducts. Inside the breast, cancer develops in the fatty tissue known as the adipose tissue and the fibrous connective tissue. Noncancerous cells that are not under control frequently infect other healthy breast tissues and then spread to lymph nodes beneath the arms. Persistent changes are experienced in the breast and should be evaluated by a physician.

The best chance of curing BC is early diagnosis, which helps detect cancerous cells to allow treatment that is less complex, less expensive, and more effective [3]. In recent years, medical imaging has made tremendous progress. Detection of breast cancer is done through mammography [4], ultrasound [5] and magnetic resonance imaging (MRI) [6]. In case of suspicious cancerous lesions, either a nodule [7] or microcalcification [8] biopsy is proposed. This biopsy confirms cancer diagnosis, stage, and type [9].

Digitized tissue samples make it easy to convert microscopic slides into histology images. Automatic segmentation of BC cells from histology images is key in analyzing cell morphology. This is important to assist diagnosis by different pathologists in oncology [10]. BC diagnosis has matured with time through multiple research methods, for instance, segmentation, detection, and classification tasks.

V. Majanga (✉) · E. Mnkandla
University of South Africa, Pretoria, South Africa
e-mail: mnkane@unisa.ac.za



I. Woungang, S. K. Dhurandher (eds.), *The 7th International Conference on Wireless, Intelligent and Distributed Environment for Communication*,
Lecture Notes on Data Engineering and Communications Technologies 237,
https://doi.org/10.1007/978-3-031-80817-3_9

Consequently, Abhisheka et al. [11] suggests that computer-aided diagnosis (CAD) is required, to assist radiologists in various analysis stages and offer a second opinion on final decisions. CAD systems combine multiple elements, for example, artificial intelligence, computer vision, and medical imaging processing. These systems help minimize the effect of operator dependency and increase diagnostic sensitivity and specificity. Identifying and classifying correctly BC subtypes is important in clinical treatment. However, it is being discovered that some of the conducted biopsies are not cancerous. Therefore, CAD systems are used to distinguish benign and malignant cancerous lesions to reduce unnecessary biopsies. These automated approaches are used to save time and reduce errors.

Different imaging modes with CAD systems are highly effective in BC fatality reduction, especially in the early stages. Artificial intelligence is greatly utilized to automate image segmentation, detection, and classification to accurately and speedily estimate risks. It assists in timely preventive measures and medical professionals make productive conclusions through deep learning in clinical tasks.

CAD systems have achieved remarkable evaluation performance in early BC diagnosis. Research by Mridha et al. [12] offers a well-systematic and analytical review of aspects such as the model architecture of BC diagnosis, image datasets, image pre-processing, ways of BC imaging, performance measurements, and research directions.

Analysis of BC histology images requires pre-processing to obtain useful information. Image segmentation is used mostly as a significant image processing procedure. This representation of BC histology images is done by identifying various ROIs from a larger image to locate cancerous cells. Image segmentation groups image into several segments to aid object and edge identification.

Breast tumor detection depends on segmentation methods. Research by Michael et al. [13] groups segmentation methods into classical segmentation, which includes region, threshold, and edge segmentation, machine learning segmentation, supervised and unsupervised; and deep learning segmentation.

Recent advances in medical image analysis, specifically deep learning, implement connected component analysis to help identify, detect, classify, and evaluate patterns. More importantly, these advances lead to exploring hierarchical feature representations learned from images. Connected component analysis (CCA) is invaluable in many image processing activities. Works by Wagner et al. [14] intimates that CCA divides an image into foreground and background pixel objects. Foreground pixels are then connected and extracted. Background pixels are connected to form boundary edges on images. CCA is used after pre-processing and morphology operations on an image.

Major limitations are identified by this study that hinder the preferable performance of BC histology segmentation techniques. BC histology images are in limited supply, contrary to what is needed for training and testing deep learning systems. Secondly, most augmented BC histology images are characterized by low contrast, blurry edge borders, and noise. Additionally, most segmentation techniques require expensive computational power that leads to limited feasibility.

This proposed method sorts out the limitations mentioned. Primarily, the limitation in BC histology images is handled by data augmentation. The available 24 BC histology images are augmented to create a dataset of 11,151 BC histology images. These augmented images are pre-processed via gray scaling and thresholding to remove noise. Further, dilation morphology accentuates image pixels from the larger image. Lastly, image segmentation is analyzed by connected components to distinguish foreground pixels from background pixels, identifying and extracting the region of interest (ROI) on the BC histology image. This proposed method is trained on 8079 images and tested on 2694 images. An experimental analysis of this proposed method is done and its performance is compared to state-of-the-art methods.

## 2 Related Works

Most DL techniques are impressive in the segmentation of medical images. Various researchers have identified, proposed, and developed different DL techniques for BC images. Our research highlights some of these researchers' work and is focused on those related to thresholding, deep convolution neural networks, and connected component analysis.

The authors in [15] proposed a hybrid approach that combines aspects of CNN and connected component analysis (CCA) to segment malignant breast lesions without pre-processing. This avoids image sharpness distortion at the early stages. They use K-means and fuzzy C-means methods to compare their results. Consequently, Li et al. [16] introduces an iterative, connected component analysis-based image artifact reduction algorithm. This uses a CNN to segment co-registered ultrasound (US) images to extract lesion locations and sizes to guide artifact reduction. They demonstrate its performance using Monte Carlo simulations on breast cancer patient images.

Punitha et al. [17] proposes an automated breast mass detection system to support radiologists in accurate diagnosis. In their pre-processing step, Gaussian filters are used, for automated detection of breast masses, optimized region growing technique is used for initial seed points, and thresholds are optimally generated using a swarm optimization method referred to as dragonfly optimization (DFO). Further, gray level co-occurrence matrix (GLCM) and generalized low-rank model (GLRM) techniques are used to extract textural features from the segmented images and fed into a feedforward neural network (FFNN) classifier that is trained using a back-propagation algorithm to classify images as either malignant or benign.

The authors in [18] provide a comprehensive analysis and exploration of various segmentation methods. These methods include classical methods, machine learning, deep learning, and manual segmentation methods, and their application in medical imaging for BC detection. Classical segmentation, including threshold-driven segmentation and edge-driven segmentation, is known for its filter use, while region-based methods achieve precise representation. A clear emphasis on creating, selecting, and comparing classical approaches guidelines. Further, machine learning segmentation includes supervised and unsupervised methods that support training models with annotated images and pathology reports. This focuses on BC segmentation efficiency tasks.

To visualize BC, radiologists prefer to use medical images such as mammograms and ultrasound images. For cancerous lesion detection, a tumor ROI is extracted from an image. This segmentation step is time-consuming with noise, low contrast, and blurriness present. Pre-processing is introduced to remove unwanted information from the image. Segmentation affects the classification of images into either malignant or benign classes. Various segmentation techniques proposed by Dabass et al. [19] are used to extract microcalcification ROI and breast lesions to remove pectoral muscles.

Kadry et al. [20] intimates implementation of joint thresholding and segmentation methods to enhance and extract breast tumor sections from 2D MRI slices. Further, a trilevel thresholding based on the slime mold algorithm and Shannon entropy (SMA+SE) is implemented to enhance breast tumor sections, and watershed segmentation is used to extract breast tumor sections. After the extraction of breast tissue sections, a study is performed between ground truth images and the breast tissue sections, and necessary image performance values are computed.

Exceptional medical segmentation of images is of great significance to improving the accuracy of lesion area detection. Multi-level threshold image segmentation framework based on novel differentiation evolution proposed by Yang et al. [21] uses 2D Kapur's entropy and 2D histogram to improve the efficiency of subsequent image analysis and diagnosis. An enhanced differential evolution is used in the framework based on a roundup search, elite levy mutation, and the decentralized foraging strategy to explore optimal thresholds.

The authors in [22] proposed an improved threshold-based segmentation model to obtain ROI. Specifically, a hybrid segmentation method is used for breast region and pectoral muscle boundaries in mammogram images based on thresholding and machine learning methods. The breast region is highlighted by eliminating bands of wavelet transform for boundary estimation. A new thresholding method is used to determine the breast boundary, and morphology operations, combined with masking, are employed to correct the overestimated boundary through deleting small objects.

An architecture referred to as Connected-UNets proposed by Baccouche et al. [23] was used to connect two UNets through additional modified skip connections. Integration of ATrous Spatial Pyramid Pooling (ASPP) in the two standard UNets is to emphasize contextual information within the encoder-decoder network architecture. Further, the proposed architecture is used on the Attention UNet (ANet) and the Residual UNet (ResUNet).

A key step in identifying abnormality in breast cancer lesions (benign from malignant) is the extraction of the ROI. Zeebaree et al. [24] proposes a new approach for breast ROI extraction. It is based on local pixel information and neural networks. It includes model training to extract several batches from both ROI and background.

It also involves image testing with a fixed-size window to detect the ROI from the background and a distance transform to identify the ROI and remove non-ROI.

Mammography is one of the best ways for BC detection in its formative stages. It provides additional information about the cancer, such as malignant, benign, or normal. Jasti et al. [25] proposes a transformative approach, for classifying and detecting BC lesions, based on image processing and machine learning. This proposed model combines image pre-processing, feature extraction, feature selection, and machine learning techniques. It helps in the classification and identification of skin diseases. Further, a geometric mean filter is used to enhance image quality, and AlexNet architecture is used for extracting features. A relief algorithm and machine learning methods such as least square support vector machine, KNN, random forest, and naive Bayes for disease categorization and detection are used to extract features.

A novel end-to-end deep learning framework proposed by Li et al. [26] for mammogram image processing computes mass segmentation and concurrently predicts diagnosis results. This method maps a coupled problem manner with important boundary and shape knowledge. It uses the Locality Preserving Learner (LPL) path to hierarchically extract and exploit built-in features of the input. Besides, it uses the other Conditional Graph Learner (CGL) path to focus on generating geometrical features via modeling pixel-wise images to mask correlations. With the integration of these two learners, the cancer semantics and representations are learned well. These component learning paths complement each other and contribute to improving mass segmentation and cancer classification problems.

Early diagnosis of BC is a critical step, and explicit segmentation of breast tumors with meticulous shape and edge boundaries is invaluable to definite diagnoses and reduction of patients' discomfort. A novel method based on U-Net proposed by Yu et al. [27] was used to improve the accuracy of tumor segmentation in breast ultrasound images. A Res Path is introduced into the U-Net to reduce the difference between feature maps, of the encoder and decoder. Consequently, to minimize feature information loss and the vanishing gradient problem, a new connection, and a dense block from the input of the feature maps in the encoding-to-decoding area are added to the proposed method.

This paper proposes connected component analysis method to segment BC histology images. It specifically groups image objects by giving them labels similar to those of their adjacent objects. This facilitates individually counting regions that have been manipulated and measuring their properties. Labeling object elements is essential, not only when dealing with large-sized images to isolate them and count an individual region. It is used when counting the number of objects in their corresponding small neighborhoods. When designing a topology-preserving removal operation, it is important to thin transformations. In 2D images, the connected components in a 3*3 neighborhood are counted. This study proposes a similar approach for segmenting BC histology images discussed meticulously in the next section.

# 3 Proposed Method

This proposed method for segmenting BC histology images uses connected component analysis and is shown in Fig. 1 and handled by the following steps.

## 3.1 Dataset

The dataset used in this study was obtained from the Kaggle repository and has 24 unsupervised hematoxylin and eosin (H&E) images. This dataset size is limited and thus inapplicable for processing by deep-learning neural networks. To solve this, data augmentation is prescribed that increases the dataset size to allow deep learning.

### 3.1.1 Dataset Augmentation

There has been an issue with a limited number of medical images in publicly available datasets, making it impossible for neural networks to derive intelligent decisions [29]. One cardinal pre-processing method that has shown to be effective in training highly discriminate deep learning models is data augmentation [28].

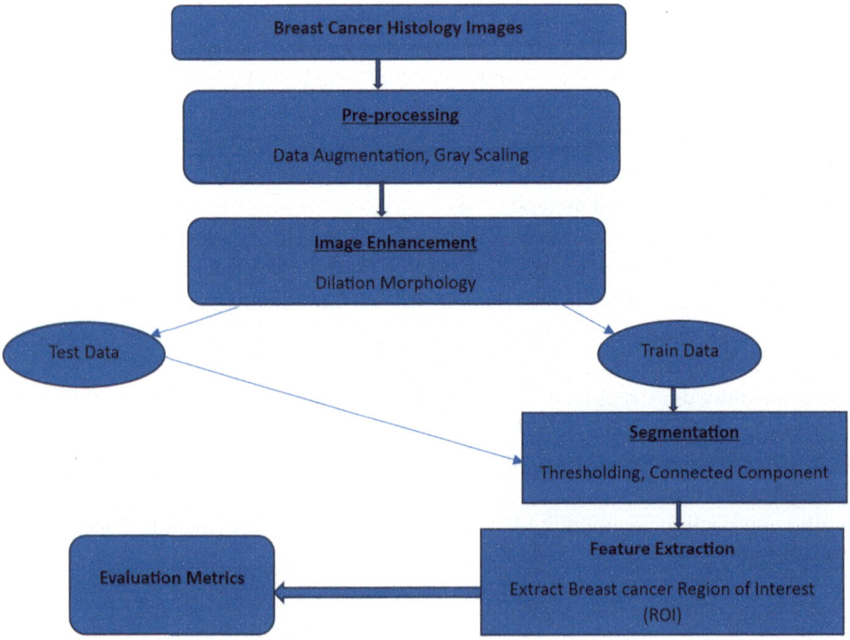


**Fig. 1** Flow diagram of the proposed segmentation method

The authors in [30] point out the essence of data augmentation as a preferred method for increasing dataset size, in specific fields with limited publicly available datasets. This is often used in medical image processing. Accessibility of large medical image datasets is a huge challenge due to ethical concerns and labeling costs. Data augmentation assists in solving this issue and greatly increases the size and variety of data available for training, without collecting new samples [31].

DL techniques require huge amounts of data, to train and test models to achieve good evaluation performance. Data augmentation is artificially creating images through different transformations to increase dataset size. Augmentation methods range from simple effective transformations such as cropping, flipping, padding, width and height shifts, rotation, flipping, and scaling.

In this proposed study, 24 unsupervised BC histology images are augmented through various transformations, namely, height and width image shifts, scaling, and rotation arguments to produce 11,151 images. Data augmentation through patch rotation and mirroring improves and increases dataset size without necessarily worsening the quality [32]. Consequently, the problem being studied is rotation invariant; this means that various pathologists can analyze BC histology images in different orientations without altering the diagnosis.

## *3.2 Image Enhancement*

Dataset preparation is done through data augmentation and image enhancement by improving contrast, brightness adjustment, and scaling to compensate for the nonuniformity caused by image illumination.

### 3.2.1 Thresholding and Morphology operations

Image enhancement is handled by gray scaling, OTSU thresholding, and dilation morphology operations.

After gray scaling and OTSU thresholding, dilation morphology diminishes foreground features while accentuating background features as shown in Fig. 2.

Dilation morphology in the image enhancement process separates BC lesions from the background. Additionally, image thresholding is used to obtain the general outline of the cancerous ROI, in the histology image, and is shown in Fig. 3.

## *3.3 Segmentation*

### 3.3.1 Connected Component Analysis

The CCA method combines pixel components with neighborhoods with the same properties. Additionally, image pixels are grouped into connected darker and

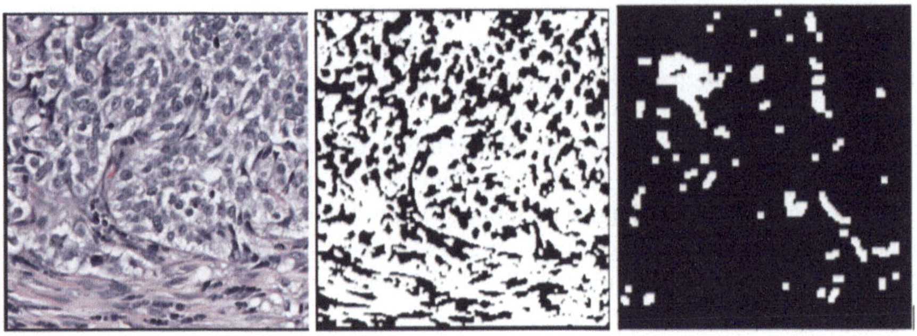

**Fig. 2** Images of original image, OTSU thresholding, and dilation morphology operations, respectively

**Fig. 3** Threshold image

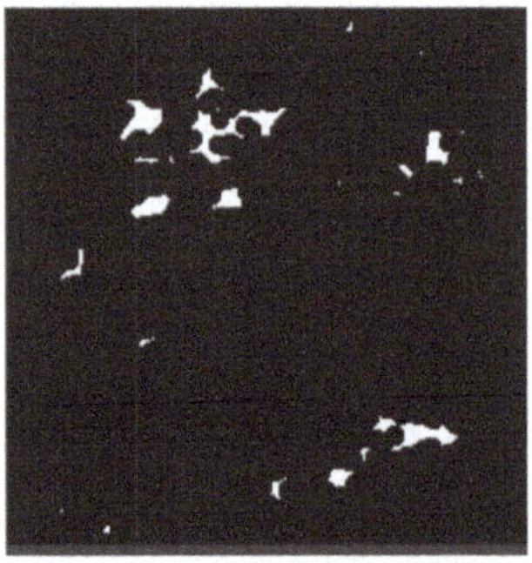

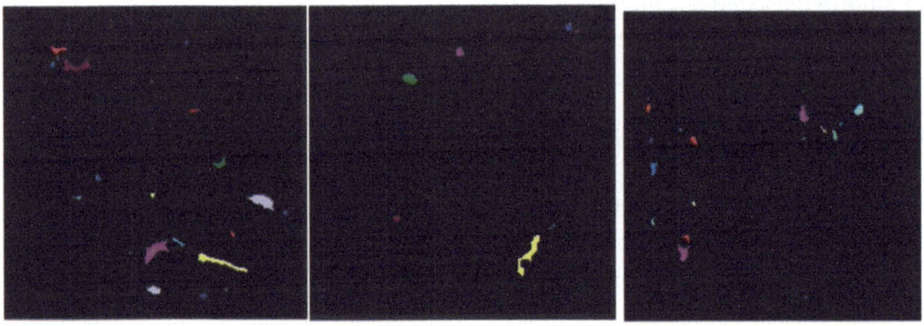

**Fig. 4** Connected component analysis images

brighter regions. Darker regions form the background, while brighter pixels form the foreground. The connected component is used to extract the ROI as shown in Fig. 4.

The detection of darker and brighter regions is influenced by the Laplacian of Gaussian (LoG) approach, a convolution kernel of the form:

$$LoG = \frac{x^2 + y^2 - 2\sigma^2}{\sigma^4} e^{-\frac{x^2 = y^2}{2\sigma^2}} \tag{1}$$

such that $\sigma$ is the kernel width. Figure 5 shows original, augmentation, thresholding, and connected component analysis images. The connected component analysis (CCA) method extracts ROI on BC histology images.

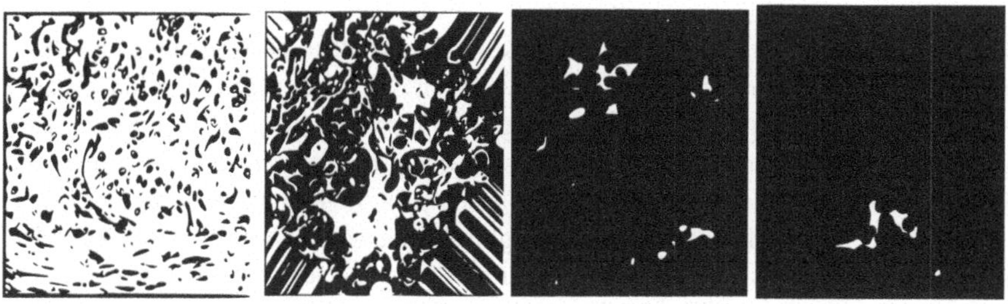

**Fig. 5** Original image, augmented image, threshold image, and connected component image, respectively

**Algorithm 1** Proposed model for breast cancer histology image segmentation

```
procedure ENCODER(Y)                                  ▷ Y : y_i is an input image with dimension (L, H).
    Initialization: Feature_i = y_0;
    for (i = 0 : N − 1) do
        AugmentedFeature_i = DataAugmentation(Feature_i);
        GrayScalingFeature_i = GrayScaling(AugmentedFeature_i);
    end for
end procedure
procedure PROCEDURE ENHANCEMENT(GrayScalingFeature_i)     ▷ It is the output of the preprocessing procedure
    for (j = 0 : M − 1) do
        DilationFeature_i = DilationMoropology(GrayScalingFeature_i);
    end for
end procedure
procedure PROCEDURE SEGMENTATION(DilationFeature_i)       ▷ It is the output of the enhancement procedure
    for (s = 0 : L − 1) do
        ThresholdingFeature_i = Thresholding(DilationFeature_i);
        ConnectedFeature_i = ConnectedComponentFeature(ThresholdingFeature_i);
    end for
end procedure
procedure PROCEDURE FEATUREEXTRACTION(ConnectedFeature_i)  ▷ It is the output of the segmentation procedure
    for (r = 0 : K − 1) do
        Feature_i = FeatureExtraction(ConnectedFeature_i);
        Segmented Image Feature= Feature_i;                  ▷ Final Segmented Output Display.
    end for
end procedure
```

## 4 Results and Discussion

Experimental results from this study are based on the analysis of the segmentation of unsupervised BC histology images using the proposed connected component analysis method. This proposed method is evaluated on augmented 11,151 BC histology images. This set of images is split into 8079 training images and 2694 testing images. Figure 1 shows an overview of the entire segmentation process specifically: dataset pre-processing, image enhancement, and segmentation. This proposed segmentation method exhibits significant results that indicate noticeable improvements in existing models. The success of this segmentation method is measured by how well it identifies and highlights cancerous lesions with their edge boundaries on BC histology images. Figure 5 shows the segmentation of cancerous lesions on unsupervised BC histology image results.

Figure 6 shows the training accuracy and training loss graphs from the proposed connected component analysis segmentation method.

Figure 7 exhibits validation accuracy and validation loss graph from the proposed connected component analysis segmentation method.

The proposed connected component analysis method to identify, highlight, and segment BC lesions through the entire unsupervised BC histology-augmented dataset led to significant results. These results are accredited to some inclusion steps, namely, dataset augmentation to increase the dataset size, image enhancement

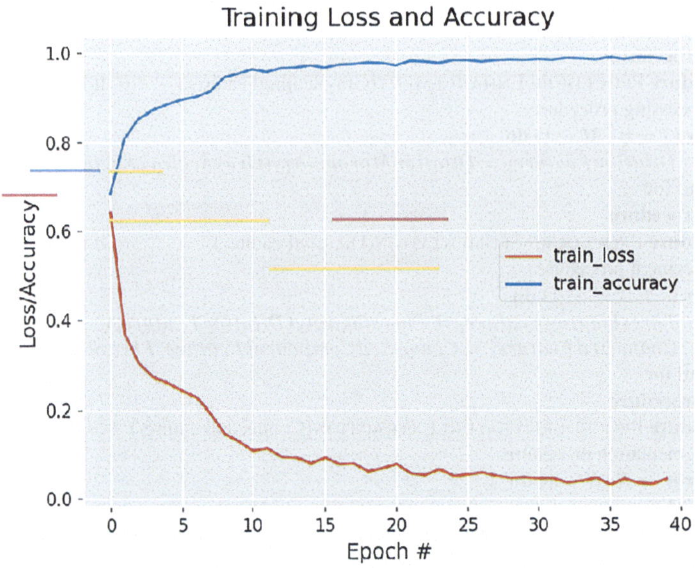


**Fig. 6** Training accuracy and training loss graphs

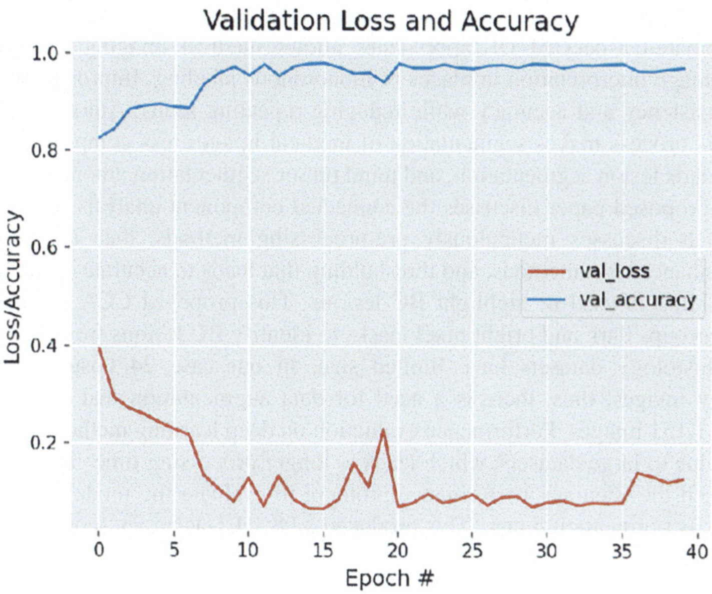


**Fig. 7** Validation accuracy and validation loss graphs

**Table 1** Overall comparison of segmentation methods

| Reference authors | Methods | Accuracy |
|---|---|---|
| Roy et al. [15] | K-means + fuzzy C-means | 90% |
| Punitha et al. [17] | Optimized region growing | 97.8% |
| Yang et al. [21] | Multi-level threshold segmentation framework | 92% |
| Majanga et al. [29] | Threshold connected component analysis | 93% |
| Zebari et al. [22] | Improved threshold-based + trained fully automated segmentation | 98.13% |
| Baccouche et al. [23] | Connected UNets | 95.88 |
| Zeebaree et al. [24] | Machine learning + region growing | 95.4% |
| Li et al. [26] | Dual convolution neural network | 93.69% |
| Proposed segmentation method | Connected component analysis) | 98% |

to improve image quality, and segmentation to extract regions of interest. Table 1 compares its performance with other state-of-the-art techniques.

## 5 Conclusion

DL techniques are invaluable in segmenting and detecting ROIs in medical images, specifically BC on histology images. Segmentation extracts regions of interest to identify objects of similar characteristics and features on an image. Moreover, the automated aspect of DL approaches allows medical imaging specialists to realize image interpretation in places of monotonous labeling. Improved efficiency aids consistency and accuracy while reducing repeating scans. Automation of the diagnosis process makes segmentation of medical images less complex and faster and this aids lesion segmentation, and mind tumor segmentation among others [33].

This proposed paper discusses the connected component analysis segmentation method. It discusses meticulously pre-processing methods, data augmentation, image enhancement methods, and thresholding that leads to accurate segmentation. Thresholding is used to highlight BC lesions. This proposed CCA segmentation method groups dark and bright pixel masks to identify BC lesions from an image.

BC histology datasets have limited size, in our case 24 unsupervised BC histology images; thus, there is a need for data augmentation that increased the size to 11,151 images. Performance evaluation on deep learning methods takes a lot of time due to large datasets, which leads to longer processing time. Therefore, our study used the accuracy performance protocol to evaluate the model's efficacy in meeting its segmentation task. This produced a 98% F1-accuracy score, which was determined by the equations below:

$$\textbf{Accuracy} = \frac{TP + TN}{TP + TN + FP + FN} \times 100 \tag{2}$$

$$\textbf{Precision} = \frac{TP}{TP + FP} \times 100 \tag{3}$$

$$\textbf{Recall} = \frac{TP}{TP + FN} \times 100 \tag{4}$$

$$\textbf{F1-score} = 2 \times \frac{\textbf{Precision} \times \textbf{Recall}}{\textbf{Precision} + \textbf{Recall}} \tag{5}$$

with TP, TN, FP, and FN representing true positives, true negatives, false positives, and false negatives, respectively.

From the literature reviewed in this study, we observed a few key areas cropping up from the segmentation and detection of BC in histology images. Some of these observations include the following:

- Most techniques employ the pre-processing step on BC histology images.
- The method used to segment BC histology images affects the performance of the model network.
- The performance of deep learning methods depends on image variability and dataset size.
- Data augmentation is key in the performance of deep learning methods, especially those with small dataset sizes.

Therefore, this paper proposes a new automatic approach to the segmentation of BC histology images. It offers a deep learning experience with large datasets through augmentation. It further makes it easier for supervised and unsupervised learning models to correctly separate and detect edge boundaries of BC lesions images in the dataset. The final model results depend heavily on image quality and weight regularization methods to help the model avoid over-fitting, thus gaining a niche over existing CAD systems. Future work looks at how to implement such optimizations. Despite the success of this proposed method, we appreciate a series of encouraging future perspectives that will enhance and improve the segmentation of BC histology images. Among them are the following:

**Data Availability and Integrity** Most DL approaches require significantly large dataset sizes for meaningful and effective performance results. Therefore, there is a need for more publicly available BC histology image datasets to assist in deep learning.

**Data Consistency** Most methods discussed in this study casually handle image pre-processing leading to loss of key global features. Some network models divide up whole slide images (WSI) into subregions; this leads to a slow learning process and high computational power. Deep learning improves this through pre-processing WSI to assist in gathering general and accurate results.

**Weight Regularization Methods** Used to improve the performance of models. This can be done through model hyper-parameter tuning such as optimizing learning rates, dropout, loss functions, activation functions, and early stopping methods.

**Compound Approaches** By combining several architectures to form a superior one to improve overall performance. This combination occurs at any stage of the model, namely, the pre-processing step, combining different model attributes to form one that will enhance training, extraction, detection, and classification of objects. Consequently, our work expands toward segmenting other image datasets not limited to breast cancer.

## References


1. K.L. Britt, J. Cuzick, K.A. Phillips, Key steps for effective breast cancer prevention. Nat. Rev. Cancer **20**(8), 417–436 (2020)
2. V.R. Allugunti, Breast cancer detection based on thermographic images using machine learning and deep learning algorithms. Int. J. Eng. Comput. Sci. **4**(1), 49–56 (2022)
3. W.R. Drioua, N. Benamrane, L. Sais, Breast cancer histopathological images segmentation using deep learning. Sensors **23**(17), 7318 (2023)
4. F. Boutaouche, N. Benamrane, Diagnosis of breast lesions using the local Chan-Vese model, hierarchical fuzzy partitioning and fuzzy decision tree induction. Iranian J. Fuzzy Syst. **14**(6), 15–40 (2017)
5. R. Huang, M. Lin, H. Dou, Z. Lin, Q. Ying, X. Jia, W. W. Xu, Z. Mei, X. Yang, Y. Dong, J. Zhou, Boundary-rendering network for breast lesion segmentation in ultrasound images. Med. Image Anal. **80**, 102478 (2022)
6. T. Ren, S. Lin, P. Huang, T.Q. Duong, Convolutional neural network of multiparametric MRI accurately detects axillary lymph node metastasis in breast cancer patients with pre neoadjuvant chemotherapy. Clin. Breast Cancer **22**(2), 170–177 (2022)

7. E. Evain, C. Raynaud, C. Ciofolo-Veit, A. Popoff, T. Caramella, P. Kbaier, C. Balleyguier, S. Harguem-Zayani, H. Dapvril, L. Ceugnart, M. Monroc, Breast nodule classification with two-dimensional ultrasound using Mask-RCNN ensemble aggregation. Diagnostic Intervent. Imaging **102**(11), 653–658 (2021)
8. R. Touami, N. Benamrane, Microcalcification detection in mammograms using particle swarm optimization and probabilistic neural network. Computación y Sistemas **25**(2), 369–379 (2021)
9. S.M. Tun, S. Alluri, V. Rastegar, P. Visintainer, W. Mertens, G. Makari-Judson, Mode of detection of second events in routine surveillance of early stage breast cancer patients. Clin. Breast Cancer **22**(7), e818–e824 (2022)
10. M.N. Gurcan, L.E. Boucheron, A. Can, A. Madabhushi, N.M. Rajpoot, B. Yener, Histopathological image analysis: a review. IEEE Rev. Biomed. Eng. **2**, 147–171 (2009)
11. B. Abhisheka, S.K. Biswas, B. Purkayastha, A comprehensive review on breast cancer detection, classification and segmentation using deep learning. Arch. Comput. Methods Eng. **30**(8), 5023–5052 (2023)
12. M.F. Mridha, M.A. Hamid, M.M. Monowar, A.J. Keya, A.Q. Ohi, M.R. Islam, J.M. Kim, A comprehensive survey on deep-learning-based breast cancer diagnosis. Cancers **13**(23), 6116 (2021)
13. E. Michael, H. Ma, H. Li, F. Kulwa, J. Li, Breast cancer segmentation methods: current status and future potentials. BioMed Res. Int. **2021**(1), 9962109 (2021)
14. T. Wagner, H.G. Lipinski, IJBlob: an ImageJ library for connected component analysis and shape analysis. J. Open Res. Softw. **1**(1), e6 (2013)
15. A. Roy, B.K. Singh, S.K. Banchhor, K. Verma, Segmentation of malignant tumours in mammogram images: a hybrid approach using convolutional neural networks and connected component analysis. Expert Syst. **39**(1), e12826 (2022)
16. S. Li, M. Zhang, Q. Zhu, Ultrasound segmentation-guided edge artifact reduction in diffuse optical tomography using connected component analysis. Biomed. Opt. Express **12**(8), 5320–5336 (2021)
17. S. Punitha, A. Amuthan, K.S. Joseph, Benign and malignant breast cancer segmentation using optimized region growing technique. Future Comput. Inform. J. **3**(2), 348–358 (2018)
18. A. Abo-El-Rejal, S. Ayman, F. Aymen, Advances in breast cancer segmentation: a comprehensive review. Acadlore Trans. AI Mach. Learn. **3**(2), 70–83 (2024)
19. J. Dabass, S. Arora, R. Vig, M. Hanmandlu, Segmentation techniques for breast cancer imaging modalities—a review, in *2019 9th International Conference on Cloud Computing, Data Science & Engineering (Confluence)* (IEEE, Piscataway, 2019), pp. 658–663
20. S. Kadry, R. Damaševičius, D. D. Taniar, V. Rajinikanth, I.A. Lawal, Extraction of tumour in breast MRI using joint thresholding and segmentation—a study, in *2021 Seventh International conference on Bio Signals, Images, and Instrumentation (ICBSII)* (IEEE, Piscataway, 2021), pp. 1–5
21. X. Yang, R. Wang, D. Zhao, F. Yu, A.A. Heidari, Z. Xu, H. Chen, A.D. Algarni, H. Elmannai, S. Xu, Multi-level threshold segmentation framework for breast cancer images using enhanced differential evolution. Biomed. Signal Process. Control **80**, 104373 (2023)
22. D.A. Zebari, D.Q. Zeebaree, A.M. Abdulazeez, H. Haron, H.N.A. Hamed, Improved threshold based and trainable fully automated segmentation for breast cancer boundary and pectoral muscle in mammogram images. IEEE Access **8**, 203097–203116 (2020)
23. A. Baccouche, B. Garcia-Zapirain, C. Castillo Olea, A.S. Elmaghraby, Connected-UNets: a deep learning architecture for breast mass segmentation. NPJ Breast Cancer **7**(1), 151 (2021)
24. D.Q. Zeebaree, H. Haron, A.M. Abdulazeez, D.A. Zebari, Machine learning and region growing for breast cancer segmentation, in *2019 International Conference on Advanced Science and Engineering (ICOASE)* (IEEE, Piscataway, 2019), pp. 88–93
25. V.D.P. Jasti, A.S. Zamani, K. Arumugam, M. Naved, H. Pallathadka, F. Sammy, A. Raghuvanshi, K. Kaliyaperumal, Computational technique based on machine learning and image processing for medical image analysis of breast cancer diagnosis. Secur. Commun. Netw. **2022**(1), 1918379 (2022)

26. H. Li, D. Chen, W.H. Nailon, M.E. Davies, D.I. Laurenson, Dual convolutional neural networks for breast mass segmentation and diagnosis in mammography. IEEE Trans. Med. Imaging **41**(1), 3–13 (2021)
27. K. Yu, S. Chen, Y. Chen, Tumor segmentation in breast ultrasound image by means of res path combined with dense connection neural network. Diagnostics **11**(9), 1565 (2021)
28. Z. Hussain, F. Gimenez, D. Yi, D. Rubin, Differential data augmentation techniques for medical imaging classification tasks, in *AMIA Annual Symposium Proceedings*, vol. 2017 (American Medical Informatics Association, 2017), p. 979
29. V. Majanga, S. Viriri, Dental images' segmentation using threshold connected component analysis. Comput. Intell. Neurosc. **2021**(1), 2921508 (2021)
30. P. Chlap, H. Min, N. Vandenberg, J. Dowling, L. Holloway, A. Haworth, A review of medical image data augmentation techniques for deep learning applications. J. Med. Imaging Radiation Oncol. **65**(5), 545–563 (2021)
31. F. Garcea, A. Serra, F. Lamberti, L. Morra, Data augmentation for medical imaging: a systematic literature review. Comput. Biol. Med. **152**, 106391 (2023)
32. T. Araújo, G. Aresta, E. Castro, J. Rouco, P. Aguiar, C. Eloy, A. Polónia, A. Campilho, Classification of breast cancer histology images using convolutional neural networks. PLoS One **12**(6), e0177544 (2017)
33. J, Mistry, Automated knowledge transfer for medical image segmentation using deep learning. J. Xidian Univ. **18**(1), 601–610 (2024)

# Optimized Smart Insole Design for Temporal Gait Parameter Measurement

**Femi Olugbon, Nozhan Ghoreishi, Dain LaRoche, Ming-Chun Huang, and Diliang Chen**

## 1 Introduction

Temporal gait parameters (TGPs) play a crucial role in analyzing human gait patterns and evaluating various clinical conditions. These parameters, which include foot contact detection, heel strike, and toe-off events, provide essential information about gait characteristics such as step duration, time spent on each foot, and walking speed [25]. Accurate detection of gait cycle onset and end is vital for analyzing kinematic and physiological signals during subsequent gait cycles [25]. By detecting foot contact, important gait parameters such as gait cycle time, swing time, stance time, single support time, and double support time can be estimated [9, 13]. TGPs have significant applications in clinical assessments and the evaluation of treatments for neurological disorders like Parkinson's disease (PD), multiple sclerosis, and cerebral palsy. PD patients commonly exhibit changes in TGPs, such as increased step time variability, with the severity of the disease correlating with the degree of variability [12, 14]. Additionally, TGPs find applications in sports, fall risk assessments, monitoring depression, activity recognition, and obesity studies [1, 2, 6, 7, 15–17, 22].

F. Olugbon · N. Ghoreishi · D. Chen (✉)
Electrical and Computer Engineering, University of New Hampshire, Durham, NH, USA
e-mail: femi.olugbon@unh.edu; nozhan.ghoreishi@unh.edu; diliang.chen@unh.edu

D. LaRoche
Department of Kinesiology, University of New Hampshire, Durham, NH, USA
e-mail: dain.laroche@unh.edu

M.-C. Huang
Data and Computational Science, Duke Kunshan University, Suzhou, China
e-mail: mh596@duke.edu



I. Woungang, S. K. Dhurandher (eds.), *The 7th International Conference on Wireless, Intelligent and Distributed Environment for Communication*,
Lecture Notes on Data Engineering and Communications Technologies 237,
https://doi.org/10.1007/978-3-031-80817-3_10

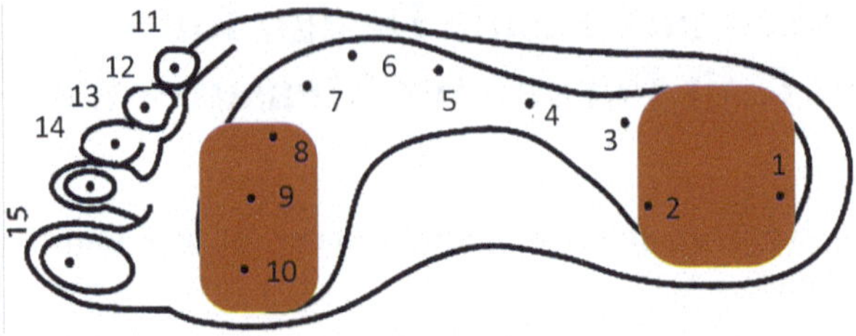


**Fig. 1** Foot anatomical regions marked with numbers showing important regions that support human body weight. The colored regions are the optimal sensor locations determined in [10]

Wearable insole sensors are commonly employed to accurately detect foot contact events and perform gait analysis. These sensors directly measure force changes during foot contact events and are commonly used for monitoring ground reaction forces (GRFs) [9]. The pressure distribution under the human foot is nonuniform, with specific anatomical points bearing most of the body weight [5]. Figure 1 shows the human foot's 15 anatomical points and the optimal sensor place for foot contact detection estimated by Chen et al. in [10]. Consequently, the accuracy of foot contact detection relies on the locations and number of pressure sensors used [23]. Previous research has proposed different numbers and locations of sensors for foot contact detection. For example, studies have utilized eight [23], six [21], five [20], four [3], and three [19] sensors, respectively.

Recent studies have proposed various sensor numbers and locations for insole design, highlighting the need for standards in pressure sensor placement for gait analysis. Optimal pressure sensor numbers and placements have been reported for different gait parameters, such as gait variability [18], center of pressure estimation [24], and foot contact detection [10]. For wearable systems, high performance, low production cost, durability, comfortability, and excellent battery life are crucial design metrics. Designing customizable insole pressure sensors that can be trimmed to fit different foot sizes is crucial to reducing costs, manufacturing overhead, and complexity while meeting the requirements of practical applications. This study proposed an optimized wearable insole measurement system that allows customization for different foot sizes and is capable of accurate foot contact detection and TGP estimation. In this study, we develop a novel optimized insole sensing system for temporal gait parameter estimation. Our proposed design uses two sensors placed in the heel and metatarsal region and utilizes a trimmable design [8], enabling the use of a single insole size adjustable to various foot sizes ranging from US 5.5 to US 11.5 male foot size.

## 2 Materials and Methods

### 2.1 Smart Insole Design

In this research, we employed a polymeric composite material called Velostat® (also known as Linqstat [4]) as the pressure-sensing element. The material, which is thin (0.1 mm), lightweight (280 g/mm$^2$), and flexible, is sandwiched between two conductive copper pads. It possesses desirable properties such as chemical and mechanical stability [11]. Figure 2 illustrates the schematic representation of the pressure sensor. As an applied force is exerted, the resistance of the pressure-sensing element changes. As depicted in the schematic diagram on the right, a resistive voltage divider mechanism (1) is employed to measure the resistance:

$$R_{sensor} = \frac{V_{sensor} \times R_{fixed}}{V_{CC} - V_{sensor}} \tag{1}$$

The placement and dimensions of the sensors were determined based on a previous study conducted by [10], which involved participants with a US 10.5 Men's foot size. A two-dimensional projection technique was utilized to establish the boundaries of the pressure sensors, catering to foot sizes ranging from 5.5 to 11.5 US men's sizes (2). The voltage across the sensing material is captured through an analog-to-digital converter and wirelessly transmitted to data acquisition software at a sampling rate of 100 Hz. An overview of the insole assembly is presented in

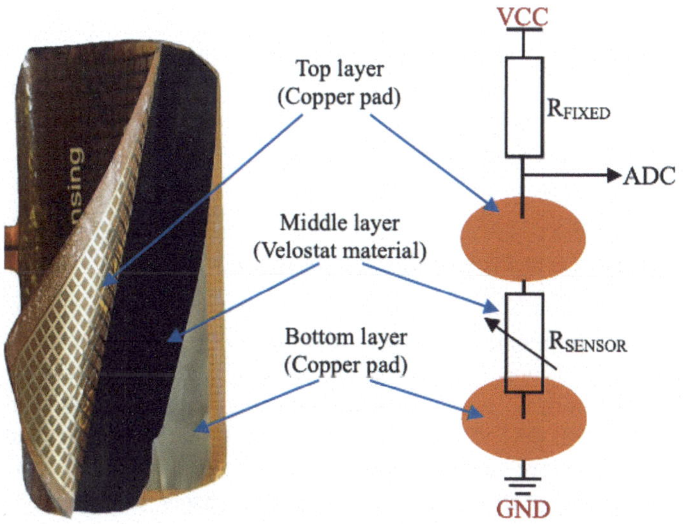


**Fig. 2** Mechanism of the pressure sensor array. The figure on the left showed a pulled-apart sensor. The schematic on the right shows the sensor mechanism

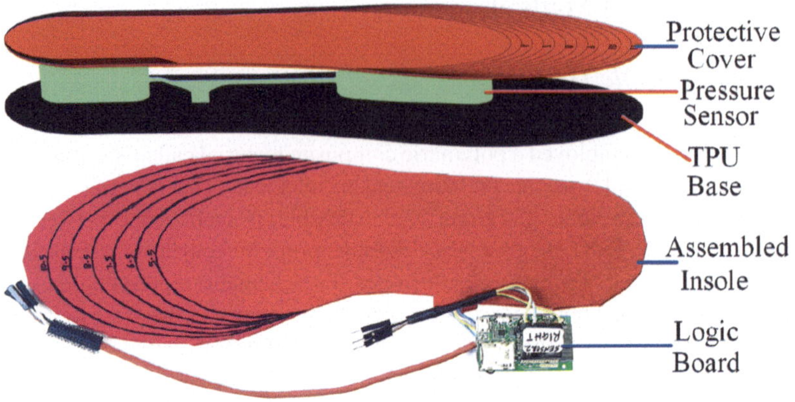


**Fig. 3** Insole assembly. The top figure shows the cross section of the sensor and the protective layers. The bottom picture shows the final assembled insole with a 1.8 mm thickness and a miniature data acquisition board (44 mm × 27 mm)

Fig. 3, where the pressure sensor is encapsulated between two flexible thermoplastic polyurethane (TPU) materials:

$$\begin{cases} x_{new} = \dfrac{L_{x\_new}}{L_{x\_base}} \times x_{base} \\ y_{new} = \dfrac{L_{y\_new}}{L_{y\_base}} \times y_{base} \end{cases} \tag{2}$$

where $x_{base}$ and $L_{x\_base}$ are the sensor coordinate in the foot breadth axis and the width of the foot for US 10.5 size and $x_{new}$ and $L_{x\_new}$ are the sensor coordinate in the foot breadth axis and the width of the foot for the projected size. The y-direction corresponds to the foot length axis.

## 2.2 *Temporal Gait Parameter Estimation*

Table 1 presents a comprehensive overview of the TGPs that can be estimated using the proposed insole system. In this section, we outline the method and algorithms employed for detecting foot contacts and calculating key parameters such as SRT, SPT, SWT, STT, SST, DST, GV, and GS. Foot contact detection is achieved through the utilization of pressure sensors that function as switches. To differentiate between the no-contact and contact phases of the pressure sensors, an implementation of an adaptive threshold (3) proposed in [9] was used. Furthermore, for reference measurements, we employed the h/p/cosmos Gaitway III instrumented treadmill (h/pcosmos Sports & Medical gmbh, Germany), which offers a proprietary set of

**Table 1** Gait parameter provided by the optimized customizable insole

| Gait parameter (ms) | Description |
|---|---|
| Stride time (SRT) | Time elapsed between two consecutive heel strikes of the same foot |
| Step time (SPT) | Time elapsed between two consecutive heel strikes of the different feet |
| Swing time (SWT) | Time elapsed between the heel strike and toe-off of the same foot |
| Stance time (STT) | Time elapsed between the heel strike and toe-off of the same foot |
| Single support time (SST) | Time duration in a gait cycle when only one foot is in contact with the ground |
| Double support time (DST) | Time duration in a gait cycle when both feet are in contact with the ground |
| Gait variability (GV) | Stride-to-stride differences in gait parameters.<br>This parameter can be estimated with all of the above gait parameters |
| Gait symmetry (GS) | A measure of similarities between gait parameters of two lower limbs.<br>This parameter can be estimated with all of the above gait parameters |

algorithms to derive all TGPs:

$$Adaptive\ Threshold = \frac{1}{n_{swing}} \sum_{k=1}^{n_{swing}} R_k + 3 * \sigma_{swing} \tag{3}$$

## 3 Results and Discussion

Three healthy participants with normal gait patterns were involved in the experiment. During the experiment, each subject wore a trimmed insole and walked on the treadmill for 2 min at speeds of 0.9 m/s, 1.2 m/s, and 1.6 m/s, respectively, with a 5-min break between trials. Data for the first and last 5 s were excluded from the analysis. For all candidate TGPs, the measurement and the differences in the measurements between the two systems were judged to be normally distributed. Accordingly, the mean and standard deviation (SD) was used for the analysis. Figure 4 and the agreement plots for the stride time parameter are shown. The results in Fig. 4 connote a strong agreement between the proposed insole design and the reference measurement system. Note that because the sampling rate is 100 Hz (resolution is 10 ms), the graph includes many overlapping points. The accuracy and precision results for all the TGPs are summarized in Table 2. The proposed insole system accurately estimates the heel strike event based on the results. An early start from swing phases was noticed due to the inaccuracy in detecting toe-off events. Figure 5 shows the normalized ground reaction force of the reference treadmill and

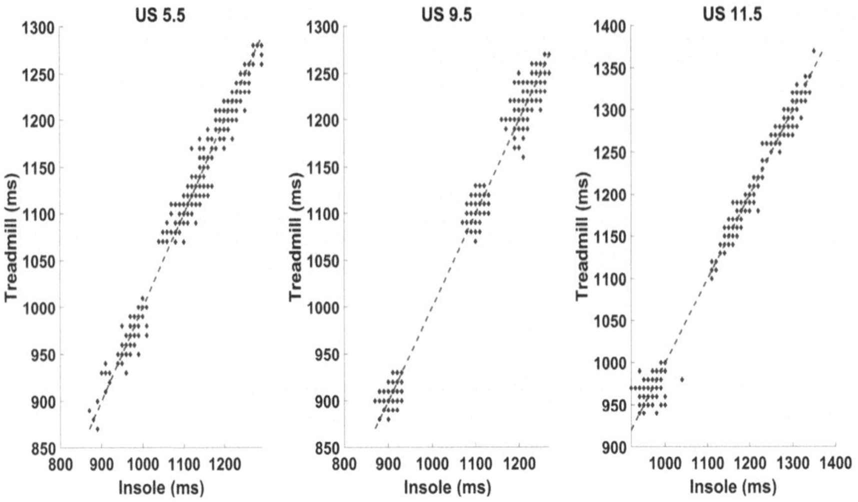


**Fig. 4** Agreement plots between the proposed insole system and reference instrumented treadmill results for stride time measured from 200 continuous effective strides

**Table 2** Accuracy and precision of TGPs for different foot sizes

| | US 5.5 | | US 9.5 | | US 11.5 | |
|---|---|---|---|---|---|---|
| TGR | MAE | SD | MAE | SD | MAE | SD |
| SRT (ms) | 8.18 | 12.31 | 9.18 | 12.75 | 7.13 | 11.08 |
| SPT (ms) | 15.08 | 13.88 | 18.64 | 22.22 | 21.89 | 17.86 |
| SWT (ms) | 15.53 | 13.96 | 18.54 | 22.29 | 21.68 | 17.86 |
| STT (ms) | 15.08 | 13.88 | 18.64 | 22.22 | 21.89 | 17.86 |
| SST (ms) | 15.58 | 13.95 | 18.45 | 22.15 | 21.34 | 17.28 |
| DST (ms) | 29.29 | 22.80 | 19.90 | 18.62 | 40.83 | 25.25 |
| HS (ms) | 6.39 | 8.87 | 7.98 | 9.70 | 5.78 | 8.11 |
| TO (ms) | 19.29 | 13.32 | 28.89 | 22.64 | 29.29 | 22.13 |

the states of the proposed insole pressure switches. The proposed insole detects toe-off events earlier than the reference system. Chen et al. in [10] reported pressure changes around the big toe during the toe-off event. Since the forefoot sensor of the proposed system only covers the forehead of the metatarsals, future works will be geared toward extending the forefoot sensor area to cover the big toe. The large error in all the TGPs is due to the errors in the toe-off detection. Improving the toe-off detection accuracy by enlarging the forefoot sensors to cover the toe region will enhance the overall precision and accuracy of the proposed system for TGP estimation.

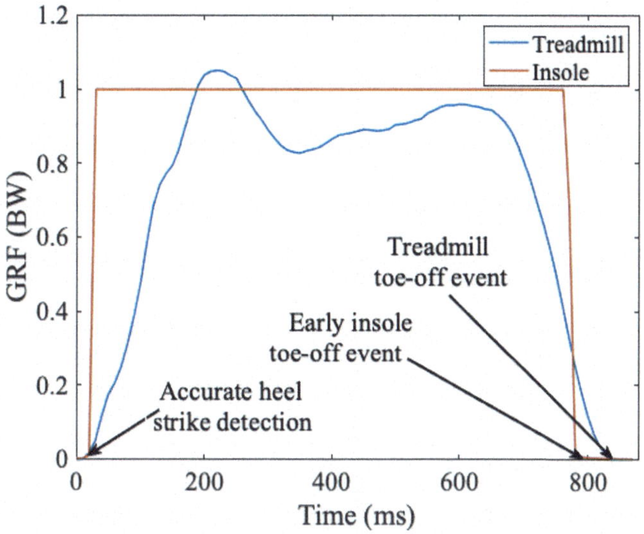


**Fig. 5** Comparison of insole and treadmill foot contact detection time. The majority of the errors in TGP estimation are due to errors in toe-off event detection. The insole pressure sensors were operated as a switch

## 4 Conclusion

In this study, we successfully developed and validated a highly efficient insole pressure sensor for estimating TGPs. The sensor is optimized and customizable, catering to foot sizes ranging from US men's 5.5 to 11.5. Eight TGPs were extracted to evaluate the performance of the optimized insole system, and the results confirmed its accuracy in estimating gait parameters. Moving forward, our future work will focus on enlarging the surface area of the forefoot sensor to cover the area under the big toe and enhancing and assessing the applicability of the proposed design for individuals with larger foot sizes and varying foot shapes (pronated and supinated), as their optimal sensor placements will likely differ.

**Acknowledgments** This research was funded by the University of New Hampshire, Collaborative Research Excellence (CoRE) Initiative, Pilot Research Partnerships (PRPs), and the UNHInnovation fund.

## References

1. S. Ansah, D. Chen, Wearable-gait-analysis-based activity recognition: A review. Int. J. Smart Sensing Intell. Syst. **15**(1) (2022). https://doi.org/doi:10.2478/ijssis-2022-0021

2. S. Ansah, F. Olugbon, S. Arthanat, D. LaRoche, D. Chen, Smart insole based shuffling detection system for improved gait analysis in Parkinson's disease, in *2023 IEEE 19th International Conference on Body Sensor Networks (BSN)* (IEEE, Piscataway, 2023), pp. 1–4
3. S.J.M. Bamberg, A.Y. Benbasat, D.M. Scarborough, D.E. Krebs, J.A. Paradiso, Gait analysis using a shoe-integrated wireless sensor system. IEEE Trans. Inform. Technol. Biomed. **12**(4), 413–423 (2008). https://doi.org/10.1109/TITB.2007.899493
4. R. Barba, Á. P. de Madrid, J.G. Boticario, Development of an inexpensive sensor network for recognition of sitting posture. Int. J. Distrib. Sensor Netw. **11**(8), 969237 (2015). https://doi.org/10.1155/2015/969237
5. P.R. Cavanagh, M.M. Rodgers, A. Liboshi, Pressure distribution under symptom-free feet during barefoot standing. Foot Ankle **7**(5), 262–278 (1987). https://doi.org/10.1177/107110078700700502
6. D. Chen, G. Asaeikheybari, H. Chen, W. Xu, M.C. Huang, Ubiquitous fall hazard identification with smart insole. IEEE J. Biomed. Health Inform. **25**(7), 2768–2776 (2020)
7. D. Chen, Y. Cai, J. Cui, J. Chen, H. Jiang, M.C. Huang, Risk factors identification and visualization for work-related musculoskeletal disorders with wearable and connected gait analytics system and kinect skeleton models. Smart Health **7**, 60–77 (2018)
8. F. Chen, Y. Cai, M.C. Huang, Customizable pressure sensor array: design and evaluation. IEEE Sensors J. **18**(15), 6337–6344 (2018). https://doi.org/10.1109/JSEN.2018.2832129
9. D. Chen, Y. Cai, X. Qian, T. Ansari, W. Xu, K.C. Chu, M.C. Huang, Bring gait lab to everyday life: gait analysis in terms of activities of daily living. IEEE Internet Things J. **7**(2), 1298–1312 (2020). https://doi.org/10.1109/JIOT.2019.2954387
10. D. Chen, N. Ghoreishi, F. Olugbon, S. Ansah, M.C. Huang, Q. Yu, Optimal pressure sensor locations in smart insoles for heel-strike and toe-off detection, in *2022 IEEE Biomedical Circuits and Systems Conference (BioCAS)* (IEEE, Piscataway, 2022), pp. 458–461. https://doi.org/10.1109/BioCAS54905.2022.9948663
11. D. Giovanelli, E. Farella et al., Force sensing resistor and evaluation of technology for wearable body pressure sensing. J. Sensors **2016** (2016). https://doi.org/10.1155/2016/9391850
12. J.M. Hausdorff, M.E. Cudkowicz, R. Firtion, J.Y. Wei, A.L. Goldberger, Gait variability and basal ganglia disorders: stride-to-stride variations of gait cycle timing in parkinson's disease and huntington's disease. Movement Disorders **13**(3), 428–437 (1998). https://doi.org/10.1002/mds.870130310. https://movementdisorders.onlinelibrary.wiley.com/doi/abs/10.1002/mds.870130310
13. W.N.W. Huang, J.M. VanSwearingen, J.S. Brach, Gait variability in older adults: observational rating validated by comparison with a computerized walkway gold standard. Phys. Therapy **88**(10), 1146–1153 (2008). https://doi.org/10.2522/ptj.20070243
14. S.M. Kim, D.H. Kim, Y. Yang, S.W. Ha, J.H. Han, Gait patterns in parkinson's disease with or without cognitive impairment. dnd **17**(2), 57–65 (2018). https://doi.org/10.12779/dnd.2018.17.2.57. http://www.e-sciencecentral.org/articles/?scid=1120132
15. P.P. Lai, A.K. Leung, A.N. Li, M. Zhang, Three-dimensional gait analysis of obese adults. Clin. Biomech. **23**, S2–S6 (2008). https://doi.org/10.1016/j.clinbiomech.2008.02.004. https://www.sciencedirect.com/science/article/pii/S0268003308000387. Research and Development on Biomechanics in China
16. M.R. Lemke, T. Wendorff, B. Mieth, K. Buhl, M. Linnemann, Spatiotemporal gait patterns during over ground locomotion in major depression compared with healthy controls. J. Psychiatric Res. **34**(4), 277–283 (2000). https://doi.org/10.1016/S0022-3956(00)00017-0. https://www.sciencedirect.com/science/article/pii/S0022395600000170
17. D. McGrath, B.R. Greene, K.J. O'Donovan, B. Caulfield, Gyroscope-based assessment of temporal gait parameters during treadmill walking and running. Sports Eng. **15**(4), 207–213 (2012). https://doi.org/10.1007/s12283-012-0093-8
18. M. Muñoz-Organero, J. Parker, L. Powell, R. Davies, S. Mawson, Sensor optimization in smart insoles for post-stroke gait asymmetries using total variation and l1 distances. IEEE Sensors J. **17**(10), 3142–3151 (2017). https://doi.org/10.1109/JSEN.2017.2686641

19. I. Pappas, T. Keller, S. Mangold, M. Popovic, V. Dietz, M. Morari, A reliable gyroscope-based gait-phase detection sensor embedded in a shoe insole. IEEE Sensors J. **4**(2), 268–274 (2004). https://doi.org/10.1109/JSEN.2004.823671
20. E.S. Sazonov, G. Fulk, J. Hill, Y. Schutz, R. Browning, Monitoring of posture allocations and activities by a shoe-based wearable sensor. IEEE Trans. Biomed. Eng. **58**(4), 983–990 (2011). https://doi.org/10.1109/TBME.2010.2046738
21. L. Shu, T. Hua, Y. Wang, Q. Li, D.D. Feng, X. Tao, In-shoe plantar pressure measurement and analysis system based on fabric pressure sensing array. IEEE Trans. Inform. Technol. Biomed. **14**(3), 767–775 (2010). https://doi.org/10.1109/TITB.2009.2038904
22. M.E. Taylor, K. Delbaere, A.S. Mikolaizak, S.R. Lord, J.C. Close, Gait parameter risk factors for falls under simple and dual task conditions in cognitively impaired older people. Gait Posture **37**(1), 126–130 (2013). https://doi.org/10.1016/j.gaitpost.2012.06.024. https://www.sciencedirect.com/science/article/pii/S096663621200241X
23. B. Wang, K.S. Rajput, W.K. Tam, A.K.H. Tung, Z. Yang, Freewalker: a smart insole for longitudinal gait analysis, in *2015 37th Annual International Conference of the IEEE Engineering in Medicine and Biology Society (EMBC)* (2015), pp. 3723–3726. https://doi.org/10.1109/EMBC.2015.7319202
24. X. Xian, Z. Zhou, G. Huang, J. Nong, C. Liu, L. Xie, Optimal sensor placement for estimation of center of plantar pressure based on the improved genetic algorithms. IEEE Sensors J. **21**(24), 28,077–28,086 (2021). https://doi.org/10.1109/JSEN.2021.3125021
25. W. Zijlstra, A.L. Hof, Assessment of spatio-temporal gait parameters from trunk accelerations during human walking. Gait Posture **18**(2), 1–10 (2003). https://doi.org/10.1016/S0966-6362(02)00190-X. https://www.sciencedirect.com/science/article/pii/S096663620200190X

# Transforming Agriculture: Disease and Pest Management Through CNN-Based Image Classification in Computer Vision

**Sree Ganesh Thottempudi, Donatien Koulla Moulla, Jelil Olatunbosun Agbo-Ajala, David Attipoe, Lateef Adesola Akinyemi, and Ernest Mnkandla**

## 1 Introduction

Using deep learning in computer vision in agriculture can transform existing processes and bring a new era of accuracy and efficiency. This thesis begins a thorough analysis of the use of deep learning techniques for computer vision in agriculture [18, 20]. The study looks at the intricacies of object detection, semantic segmentation [9], and image classification to improve crop monitoring and precision farming [22]. Computer vision, with its unparalleled accuracy and automation, has the potential to alter agriculture by removing the constraints of human observation. As we go deeper into deep learning, this study aims to emphasize the intricacies of its influence on precision farming and crop monitoring, imagining a future in which technology not only monitors but also improves agricultural foundations.

S. G. Thottempudi (✉)
Centre for Augmented Intelligence and Data Science, School of Computing, University of South Africa, Johannesburg, South Africa

Department of Electrical and Smart Systems Engineering, School of Computing, University of South Africa, Johannesburg, South Africa

D. K. Moulla · J. O. Agbo-Ajala · D. Attipoe · L. A. Akinyemi · E. Mnkandla
Centre for Augmented Intelligence and Data Science, School of Computing, University of South Africa, Johannesburg, South Africa



I. Woungang, S. K. Dhurandher (eds.), *The 7th International Conference on Wireless, Intelligent and Distributed Environment for Communication*,
Lecture Notes on Data Engineering and Communications Technologies 237,
https://doi.org/10.1007/978-3-031-80817-3_11

### 1.1 Problem Statement

Optimal crop health, efficient resource use, and yield prediction remain critical concerns in modern agriculture. Traditional crop monitoring techniques rely on manual observation, which can be inaccurate, cause delays in issue response, and are not scalable. Precision farming has challenges in accurately focusing treatments owing to a lack of real-time data and analysis.

This study examines the essential issue of inefficiencies in traditional agricultural systems, proposing that computer vision combined with deep learning can provide answers to these pressing difficulties [13]. The lack of automated, real-time monitoring technology prevents timely response to agricultural irregularities, jeopardizing resource allocation and production. Deep learning algorithms might help solve the problem by making it more straightforward to identify pests, illnesses, and crop conditions quickly and precisely.

So, the main issue is that traditional crop monitoring and precision farming in agriculture are insufficient. What is needed is a cutting-edge system that uses deep learning to deliver data-driven insights in real time.

### 1.2 Scope of Research

Here we are trying to answer the research questions. (1) How can deep learning models use visual data to identify crop illnesses, pests, and irregularities in agricultural settings [8, 12]]? (2) How scalable and adaptable are the developed deep learning models to a variety of agricultural situations, including crop types, environmental factors, and farming techniques [6]? (3) What impact do implementing deep learning-based solutions have on agricultural productivity, resource efficiency, and farming practice sustainability [16, 23]? (4) How do deep learning models deal with the complexities of agricultural data, such as changing imaging conditions, crop types, and the presence of both beneficial and harmful critters in the ecosystem [6, 12, 17]?

### 1.3 Literature Review

Vincent Mbandu Ochango, Geoffrey Mariga Wambugu, and John Gichuki Ndia investigate the effectiveness of Histogram of Oriented Gradients (HOG) in extracting features from images of maize leaves affected by diseases such as common rust, leaf spot, and northern leaf blight. The retrieved HOG characteristics were integrated with a variety of machine-learning classifiers to detect the presence of illnesses [1, 9]. The results showed that HOG features performed well when combined with certain classifiers (most likely random forest or similar), reaching considerable accuracy, precision, and recall in the classification test. This implies that HOG is a useful feature extraction approach for automated maize leaf disease detection, which might help farmers make early interventions and improve crop health management [3, 11], investigates the use of Bayesian optimization to fine-tune hyperparameters within a deep convolutional neural network (DCNN) model, specifically the AlexNet architecture. The objective was to increase the accuracy of maize (corn) disease identification using the PlantVillage dataset. The optimized DCNN model showed tremendous improvement, with an amazing 96.05% accuracy in properly recognizing several maize illnesses. TinyML Approach Towards Disease Detection. The authors [10] provide a fresh dataset of healthy and sick cashew leaves and fruits that have been augmented using RGB colour conversions, photometric/geometric augmentations, and the RaLSGAN picture-generating approach. The authors use transfer learning and a TinyML technique to increase illness identification accuracy, particularly in the circumstances with little data. Their effort is to promote early intervention and reduce crop losses in cashew agriculture [4, 15].

Karthik et al. present an attention-embedded residual convolutional neural network (CNN) for analysing tomato leaf illnesses in their study 'Attention Embedded Residual CNN for Disease Detection in Tomato Leaves'. Their model uses attention processes to concentrate on disease-relevant variables, obtaining a validation accuracy of 98% on the PlantVillage dataset and confirming the efficacy of attention-based techniques in plant disease diagnosis [14]. Subodh Bansal and Anuj Kumar provide a complete examination of how deep learning frameworks are used to identify illnesses in maize crops. They investigate different deep-learning approaches, training datasets, and the accuracy and timeliness with which these systems diagnose illnesses. Their paper highlights deep learning's potential to enhance maize disease detection when compared to conventional approaches, as well as current research trends and future initiatives in this subject [10, 21]. Based on this motivation we are proposing our methodology.

## 2 Methodology

The focus of this study is the development of two tailored convolutional neural network (CNN) models. These models were primarily designed to identify and categorize diseases in four crops: tomato, maize, cashew, and cassava. The decision to build new models rather than relying on pre-existing architectures was motivated by the specific needs of agricultural image data and the diseases that affect these crops. Below is the model architecture.

### *2.1 Model Building*

The described model uses a deep convolutional neural network (CNN) architecture inspired by the inception architecture shown in Fig. 1. This technique was chosen due to its excellent performance in picture classification challenges and its ability to

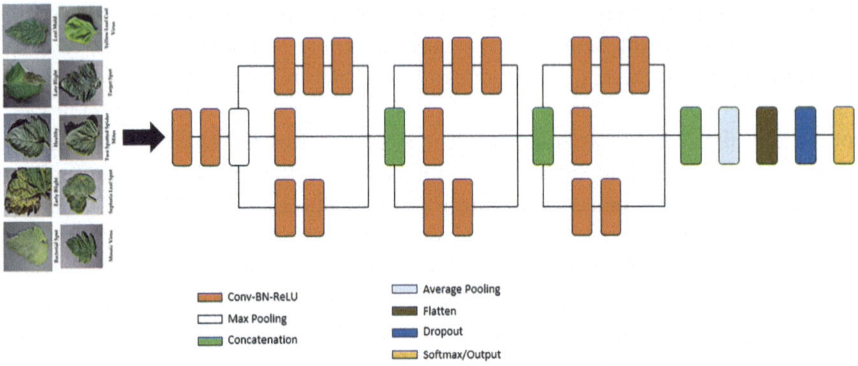


**Fig. 1** Model architecture

handle images with high dimensionality. Let us break down the architecture into its key components.

1. *Image Preparation:* The model accepts 299 × 299 RGB images (input_img).
2. *Initial Convolutional Layers:* The model starts with a series of convolutional layers (Conv2D) responsible for extracting low-level features from the input image.
   (a) Batch normalization (BatchNormalization) layers are included to stabilize training and accelerate convergence.
   (b) ReLU (Activation('relu')) activation functions introduce non-linearity, improving the model's ability to learn complex patterns.
3. *Inception Modules:* Multi-scale feature capture.
   (a) *Parallel Paths:* The inception module (mixed0, mixed1, etc.) is the foundation of the design. These modules handle an input using parallel convolutions with different kernel sizes (e.g., 1 × 1, 3 × 3, 5 × 5) and average pooling. This retrieves patterns across various scales at once.
   (b) *Dimensionality Reduction:* 1 × 1 convolutions are used strategically to reduce dimensionality before computationally expensive larger-kernel convolutions, which improves efficiency.
   (c) *Concatenation:* The outputs of all parallel routes are concatenated (Concatenate), giving the network full multi-scale feature representations.
4. *Feature Reduction and Classification:*
   (a) *Global Average Pooling:* Near the conclusion of the network, a global average pooling layer (AveragePooling2D) replaces completely linked layers. This considerably decreases the number of parameters, making the model less likely to overfit.
   (b) *Dropout layers (Dropout):* Randomly lose a fraction of connections during training, which contributes to model regularization and prevents overfitting.

(c) *Dense Layers:* Densely linked layers (Dense) process flattened features from global average pooling before creating class probabilities with a softmax activation (Dense (num_classes, activation = 'softmax').

5. *Rationale for Design Choices.*

(a) *Inception Modules:* Using inception modules has various benefits:

- Enhanced representational power through multi-scale feature extraction.
- Computational efficiency is realized by reducing dimensionality using $1 \times 1$ convolutions.

(b) *Batch Normalization: Accelerates training, minimizes sensitivity to initialization, and serves as a sort of regularization.*
(c) *Global Average Pooling: Replaces typical fully connected layers, making the model more resistant to spatial translations and less susceptible to overfitting.*
(d) *Dropout: A powerful regularization strategy that reduces overfitting and improves generalization.*

6. *Key Design Principles.*

(a) *Multi-Scale Processing:* Inception modules allow the model to extract features at diverse sizes, which is critical for picture recognition because relevant information may reside at different levels [19].
(b) *Computational Efficiency:* $1 \times 1$ convolutions inside inception modules reduce dimensionality, making the model computationally efficient regardless of depth.
(c) *Robustness:* Batch normalization, dropout, and global average pooling work together to improve model generalization while preventing overfitting (Table 1).

## 2.2 Sequential Development of Model

The research adventure began with the development of the first convolutional neural network (CNN) model. This original model was a pioneering step in investigating CNN capabilities for disease classification across four essential crops: tomato, maize, cashew, and cassava. However, upon review, the model performed anywhere from below average to mediocre. Recognizing the possibility for improvement, another CNN model was created. This second model was not just an iteration but an improved version of its predecessor [2]. The major focus during development was on minimizing the model's complexity.

**Table 1** Layers in the model

| Layer type | Filter size/stride | Output shape | Other parameters |
|---|---|---|---|
| Input | | 299 × 299 × 3 | – |
| Conv2D | 3 × 3/2 | 149 × 149 × 32 | Padding = 'valid', use_bias = False |
| BatchNormalization | | 149 × 149 × 32 | axis = 3 |
| Activation (ReLU) | | 149 × 149 × 32 | ReLU |
| Conv2D | 3 × 3/1 | 147 × 147 × 32 | Padding = 'valid', use_bias = False |
| BatchNormalization | | 147 × 147 × 32 | axis = 3 |
| Activation (ReLU) | | 147 × 147 × 32 | ReLU |
| Conv2D | 3 × 3/1 | 147 × 147 × 64 | Padding = 'same', use_bias = False |
| BatchNormalization | | 147 × 147 × 64 | axis = 3 |
| Activation (ReLU) | | 147 × 147 × 64 | ReLU |
| MaxPooling2D | 3 × 3/2 | 73 × 73 × 64 | – |
| Conv2D | 1 × 1/1 | 73 × 73 × 80 | Padding = 'valid', use_bias = False |
| BatchNormalization | | 73 × 73 × 80 | axis = 3 |
| Activation (ReLU) | | 73 × 73 × 80 | ReLU |
| Conv2D | 3 × 3/1 | 71 × 71 × 192 | Padding = 'valid', use_bias = False |
| BatchNormalization | | 71 × 71 × 192 | axis = 3 |
| Activation (ReLU) | | 71 × 71 × 192 | ReLU |
| MaxPooling2D | 3 × 3/2 | 35 × 35 × 192 | – |
| Inception Module (mixed0) | | 35 × 35 × 256 | Various configurations |
| Inception Module (mixed1) | | 35 × 35 × 256 | Various configurations |
| Inception Module (mixed2) | | 35 × 35 × 256 | Various configurations |
| AveragePooling2D | 7 × 7 | 5 × 5 × 256 | – |
| Flatten | | 6400 | – |
| Dropout | | 6400 | rate = 0.4 |
| Dense | | 2048 | ReLU |
| Dropout | | 2048 | rate = 0.4 |
| Dense (Output) | | 18 | Softmax |

## 2.3 *Data Collection*

### 2.3.1 Dataset Overview

The dataset, obtained from Mendeley (Dataset Link), was divided into two sections: raw and enhanced photos. The basic dataset had 24,881 photos from the four key crops, but the enhanced dataset contained 102,976 images (25,811-Cashew, 26,330-Cassava, 23,657-Maize, and 27,178-Tomato), divided into 22 groups [2, 5, 10, 18].

### 2.3.2 Crop-Specific Data and Classes

Each crop in the dataset had its own set of classifications that corresponded to different disease statuses and healthy conditions. For example, cashew has five classes: anthracnose, gummosis, healthy, leaf miner, and red rust. Cassava has five classifications: bacterial blight, brown spot, green mite, healthy, and mosaic. Maize has seven classifications: autumn armyworm, grasshopper, healthy, leaf beetle, leaf blight, leaf spot, and stripe virus. Tomatoes were also classified into five classes: healthy, leaf blight, leaf curl, Septoria leaf spot, and Verticillium wilt.

## *2.4 Data Pre-processing*

### 2.4.1 Augmented Data Selection

For this investigation, just the expanded component of the dataset was used. The choice to omit the raw data was based on the enhanced dataset's adequacy and diversity, which supplied enough data for efficient model training without the need for further augmentation.

### 2.4.2 Class Consolidation and Image Resizing

Originally, the dataset had 22 classes [12]. However, to expedite the categorization procedure and to account for the overlap of some circumstances between crops, the number of classes was lowered to 18. This consolidation entailed grouping related subcategories, such as 'leaf blight', which occurs in both tomatoes and maize, into a single category, as well as 'healthy', which appears in all four crops. That brings it down to 18 classifications. Furthermore, all photos were scaled to a consistent dimension of $299 \times 299$ pixels, down from their original size of $400 \times 400$.

### 2.4.3 Data Augmentation and Splitting

The data are already divided into train and test folders, each with 18 classes. The data preparation involves rescaling photos using Keras' ImageDataGenerator with a rescaling factor of validation split was used to guarantee that a portion of the dataset was saved for model validation. The dataset is now partitioned into three sets: training, validation, and testing. The exact distribution of photos throughout these sets, including the number of images per class in both the training and testing sets, will be properly maintained.

#### 2.4.4 Pre-processing Parameters

Pre-processing settings were carefully selected to improve model training and performance [3, 7]. They included:

- Number of Classes: 18,
- Batch Size: 32,
- Target Image Size: 299 × 299 pixels,
- Class Names: ('healthy', 'leaf_blight', 'leaf_curl', 'septoria_leaf_spot', 'verticulium_wilt', 'streak_virus', 'leaf_spot', 'leaf_beetle', 'grasshopper', 'fall_armyworm', 'mosaic', 'green_mite', 'brown_spot', 'bacterial_blight', 'red_rust', 'leaf_miner', 'gummosis', 'anthracnose'). Data generators were built to facilitate training, validation, and testing.

#### 2.4.5 Model Compilation and Learning Rate

The training of the custom CNN model began with its compilation using the Adam optimizer. The initial learning rate was set at 0.001, a common beginning value that strikes a compromise between rapid learning and avoiding possible minima in the loss landscape [1]. The model's loss function was defined as 'categorical_crossentropy', which is appropriate for multi-class classification tasks, and accuracy was chosen as the criterion for performance.

#### 2.4.6 Learning Rate Adjustment and Training Epochs

Implementing the ReduceLROnPlateau callback was an important part of the training process. This function changed the learning rate based on the validation loss. Specifically, if the validation loss did not improve after five epochs ('patience'), the learning rate was cut by 0.1, with a lower constraint of 0.00001. This strategy aids in fine-tuning the model by taking smaller steps as it approaches a minimum, reducing overshooting.

The model was trained for 30 epochs, which was judged sufficient to ensure convergence of loss and accuracy measures without significant overfitting. The training method entailed iterating over the training dataset in batches, as described by the train_generator, while also verifying the model on a different dataset using the validation_generator.

### *2.5 Validation*

#### 2.5.1 Integration of Validation in the Model Training Process

In this work, validation was not only a checkpoint, but it was also an essential aspect of the model training process. The addition of a validation set to the training set allowed for continuous monitoring of the model's performance on data that had not been trained on, offering real-time insights regarding its generalization capabilities.

### 2.5.2 Early Stopping Mechanism

To avoid overfitting, an early halting mechanism was introduced. The training process was rigorously examined for indicators of plateauing or a drop in validation accuracy. If the model's performance on the validation set did not increase after a certain number of epochs, the training was stopped to ensure that the model preserved its generalization capacity and did not adhere to the training data.

## *2.6 Testing Approach*

For testing, the model's performance was assessed using a different test dataset. The model's evaluate function on the test generator was used to evaluate the model using previously unknown data. The test dataset was handled in batches, much as training and validation.

### 2.6.1 Detailed Performance Metrics

In addition to accuracy, a variety of performance indicators were used to offer a comprehensive understanding of the models' capabilities. Precision, recall, and F1 scores were provided for each class. These measurements provided a more detailed insight into the models' strengths and faults, particularly in dealing with unbalanced classes.

### 2.6.2 Confusion Matrix Analysis

Confusion matrices for each model were created to help visualize the various sorts of mistakes produced. This research proved useful in detecting any systematic misclassifications and providing insights into the models' prediction behaviour. By thoroughly inspecting the confusion matrices, options for further model refining were suggested.

### 2.6.3 Receiver Operating Characteristic (ROC) Curves

ROC curves for each class were used to assess the models' diagnostic performance. The area under the ROC curve (AUC) provides a unified assessment of overall performance across many threshold settings. This research was particularly beneficial in determining the models' ability to differentiate across classes.

## 2.7 *Transfer Learning Approach with Fine-Tuning*

The use of transfer learning techniques was an important feature of this study. Both bespoke models were modified to diagnose illnesses in an additional range of crops, including potato, grape, apple, pepper bell, and maize. This stage was critical for assessing the models' adaptability and performance in larger agricultural situations.

### 2.7.1 Comparative Analysis with Pre-trained Models

The bespoke models' performance was compared to numerous pre-trained models from the Keras library, including InceptionV3, ResNet50, VGG16, MobileNet, InceptionResNetV2, EfficientNetB7, Xception, DenseNet201, and ResNet152. This comparison offered a thorough grasp of how bespoke models compare to well-established architectures in the field of deep learning.

### 2.7.2 Model Fine-Tuning

From the perspective of improving the performance of deep learning models on specific tasks, fine-tuning pre-trained models is an effective method. Using the information stored in models pre-trained on massive datasets may adjust these models to target goals with relatively little data. This method not only speeds up the training process but also increases model generalizability.

### 2.7.3 Model Selection and Modification

Our technique begins with the selection of an appropriate base model that has been pre-trained on a broad dataset. The base model selection is essential, depending on the closeness of the pre-training tasks to our goal task. For the sake of the experiment, the architecture of this base model was tweaked to meet the categorization requirement.

The modification procedure entails truncating the basic model by eliminating the top layers, hence lowering its complexity and relevance to the original purpose. It used a Global Average Pooling 2D (GAP2D) layer to compress the feature maps into a single vector per map, decreasing overfitting by lowering the overall number of parameters. Following that, a dense layer with 1024 neurons and ReLU activation provides non-linearity, allowing for the detection of complicated patterns in the data. The last layer, a dense layer with a softmax activation function, is adjusted to the number of classes in our dataset and generates probabilities for each one.

### 2.7.4 Training Procedure

The training method entails freezing the weights of the base model's original layers in order to preserve learned features. For each layer in base_model.layers, set layer.trainable = False to freeze the weights of the pre-trained model. Initially, you should focus on training only the additional categorization layers that have been put on top. Only the weights of newly added layers are changed during training. The improved model is compiled using the Stochastic Gradient Descent (SGD) optimizer, with a learning rate of 0.001 and a momentum of 0.9. The model is trained to minimize categorical cross-entropy loss, using accuracy as a performance indicator.

The model batches were fed with training data via data generators, with the number of steps per epoch determined by batch size and total number of training samples. The validation data undergo comparable validation stages. This approach enables efficient memory use and real-time data augmentation.

# 3 Results

In this section, we describe the results of a series of comparison studies conducted on two custom-developed machine learning models, referred to as Model 1 and Model 2. The evaluation was carried out across different crop datasets, with a special emphasis on the models' capability to generalize their acquired patterns via transfer learning. The datasets covered a variety of crop types, such as grapes, pepper bells, maize, and potatoes, each with its own set of characteristics and problems.

## *3.1 Architecture Comparison—Similarities*

- **Inception-Inspired:** Both models use the main concept of inception modules, parallel convolutions at various sizes, followed by concatenation.
- **Overall Structure:** The overall flow is similar: early convolutions, repeating inception modules, and final classification layers.

## *3.2 Architecture Comparison—Key Differences*

- **Complexity Inside Inception Modules:** Model 1's modules appear to be more complicated, with an average pooling branch alongside the convolutional branches. Model 2 often employs fewer convolutions per module.
- Model 1 has a higher parameter count due to the additional paths in its modules.

## 3.3 Training and Validation

### 3.3.1 Potential Effects of the Differences

Model 1 could capture more detailed and multi-scale characteristics but may be more prone to overfitting. Model 2, while computationally quicker and marginally less prone to overfitting, may underperform if the task demands particularly sophisticated pattern identification. Figure 2 explains the training accuracy and validation accuracy and Fig. 3 explains the loss and validation loss of Model 1 and Model 2 combination. Figure 4 explains the support for Model 1 and Model 2.

### 3.3.2 Observations

Model 1 shows more overfitting. As training advances, a rising disparity is seen between training and validation accuracy. Model 2 has overfitting potential, although to a lesser extent. Model 2 appears to converge faster. Its first validation loss is lower than that of Model 1. Interestingly, Model 1 has a somewhat better peak validation accuracy (0.8476 around epoch 80) than Model 2 (0.8357 around epoch 190). Model 2 exhibits somewhat less variation in validation loss and accuracy across epochs.

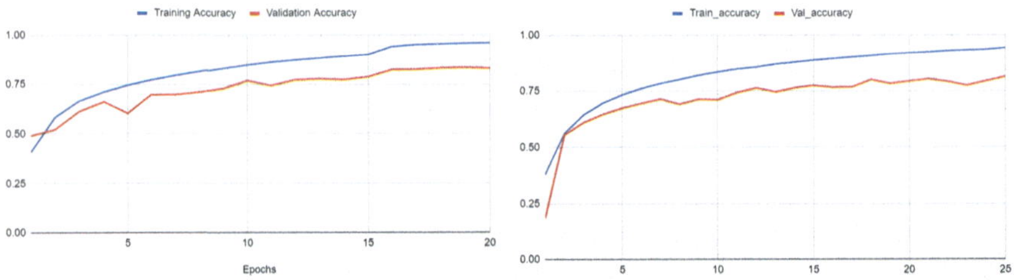


**Fig. 2** Models 1 and 2—training accuracy versus validation accuracy

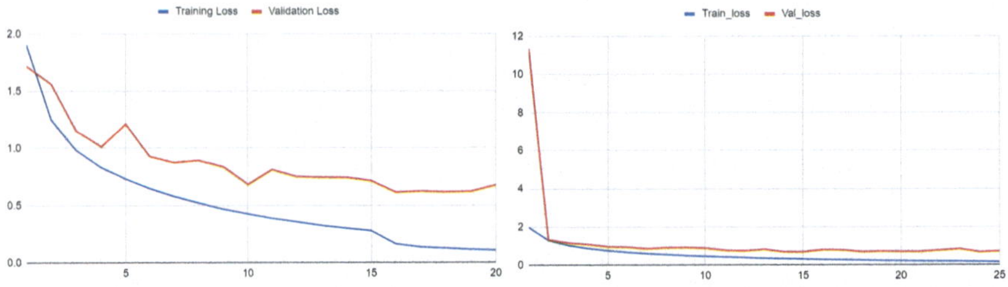


**Fig. 3** Models 1 and 2—training loss versus validation loss

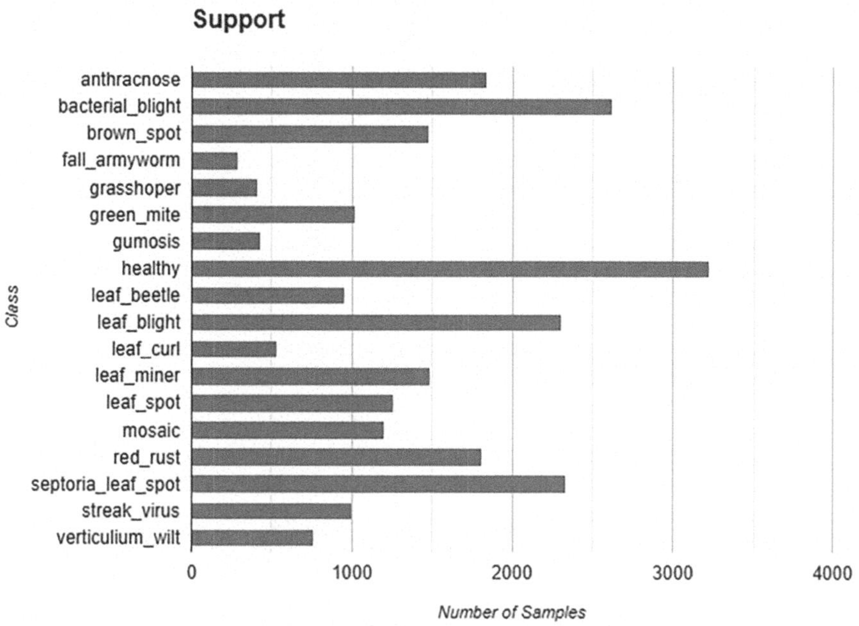


**Fig. 4** Support for both Model 1 and Model 2

Model 1's Complexity: Its better ability to learn complicated patterns appears to help it attain higher peak validation accuracy, but it also leads to more overfitting.

Model 2's Simplicity: Because of its relative simplicity, it is more resistant to overfitting, has faster convergence, and delivers significantly smoother training. However, it may lack the representational capability for the most complex characteristics required for optimal performance.

## *3.4 Testing and Validation*

### 3.4.1 Potential Effects of the Differences

Accuracy: Both models perform well on the test set (Model 1: 91%, Model 2: 92%). Model 2 has a little advantage Figs. 5 and 6 explain that. Tables 2 and 3 can demonstrate the accuracy.

Macro-Averages: Similar macro-averages for accuracy, recall, and F1 scores indicate that both models have broadly balanced performance across classes.

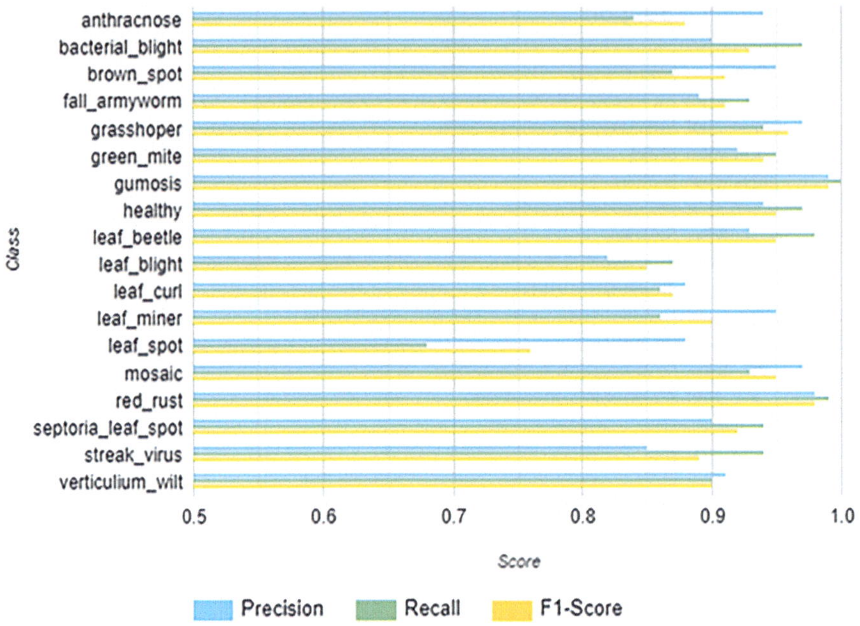


**Fig. 5** Precision, recall, F1 score for Model 1

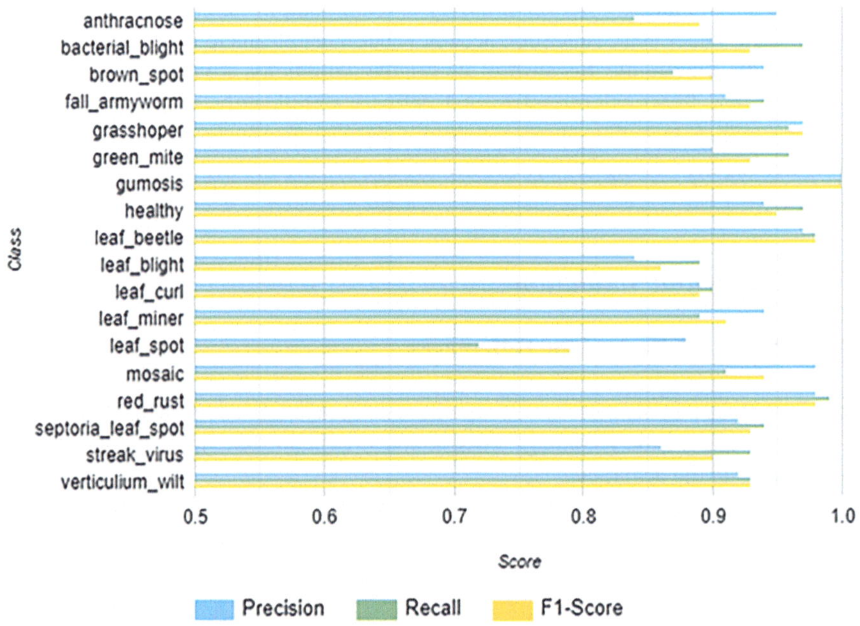


**Fig. 6** Precision, recall, F1 score for Model 2

**Table 2** Accuracy for Model 1

| | Precision | Recall | F1 score | Support |
|---|---|---|---|---|
| Accuracy | | | 0.91 | 24,981 |
| Macro-average | 0.92 | 0.91 | 0.91 | 24,981 |
| Weighted average | 0.92 | 0.91 | 0.91 | 24,981 |

**Table 3** Accuracy for Model 2

| | Precision | Recall | F1 score | Support |
|---|---|---|---|---|
| Accuracy | | | 0.92 | 24,981 |
| Macro-average | 0.93 | 0.92 | 0.92 | 24,981 |
| Weighted average | 0.92 | 0.92 | 0.92 | 24,981 |

### 3.4.2 Observations

Considering the accuracy, both models perform well on the test set (Model 1: 91%, Model 2: 92%). Model 2 has a little advantage.

Macro-Averages: Similar macro averages for accuracy, recall, and F1 scores indicate that both models have broadly balanced performance across classes. Model 1 struggles more with leafspot (poor recall, F1 score) than Model 2 and scores on verticullium_wilt are somewhat lower than in Model 2. These are some of the weak points of Model 1.

Recall for anthracnose is somewhat lower in Model 1; this is the weak point in Model 2. Both models perform admirably, although Model 2 has a slight advantage in overall accuracy and class-specific measures.

## *3.5 Confusion Matrix*

### 3.5.1 Potential Effects of the Differences

The strong points in Model 1 are it performs remarkably well in classes such as bacterial blight, healthy, and septoria_leaf_spot, with extremely high true positive rates. It also produces fewer false positives in most classifications, indicating high specificity and the weaknesses are leaf_blight (misclassified as leaf_curl and leaf_miner), leaf_curl (misclassified as leaf_blight and leaf_miner), and leaf_miner (misclassified as anthracnose and leaf_blight), all having significant misclassifications.

The strong points in Model 2 are that it has high true positive rates for bacterial_blight, healthy, and septoria_leaf_spot. Figure 7 explains that. There is a modest improvement in accurately recognizing leaf_blight when compared to Model 1 and the weaknesses in Model 2 are that it exhibits greater uncertainty in the leaf_miner category, resulting in more false positives when predicting leaf_blight and leaf_curl. In addition, it has a larger anthracnose misclassification rate than Model 1.

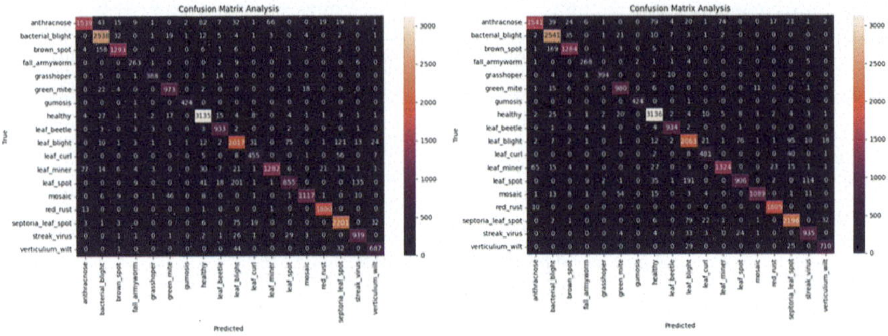


Fig. 7 Confusion matrix for Models 1 and 2

### 3.5.2 Observations

The two models perform similarly, with high true positive rates in the same classes and comparable misclassification patterns. Model 1 appears to have a minor advantage in recognizing leaf_miner and anthracnose accurately. Model 2 improves marginally in identifying leaf_blight but is less accurate in distinguishing between leaf_miner, leaf_curl, and anthracnose. Both models struggle with comparable classes, which might imply that these classes are fundamentally more difficult to differentiate, either due to similarities in the characteristics of certain plant diseases or a smaller sample size in the training data.

## *3.6 Confusion Matrix*

### 3.6.1 Potential Effects of the Differences

From Fig. 8 In Model 1 anthracnose has an AUC of 0.99, which is excellent, and bacterial blight has an AUC of 1.00, indicating flawless classification. Brown spot has an AUC of 1.00, indicating flawless performance. All other classes have AUCs of 1.00 or extremely close to it, suggesting remarkable overall performance. In Model 2, anthracnose has a little lower AUC of 0.99, but it is still great. Bacterial blight maintains a flawless AUC of 1.00. Brown spot also has an AUC of 1.00, showing that Model 1's performance remains unchanged. All other classes perform similarly to Model 1, with AUCs of 1.00 or near-perfect results.

### 3.6.2 Observations

Both models have excellent classification ability in most classes, with perfect or near-perfect AUC ratings. Model 1 outperforms Model 2 in anthracnose categoriza-

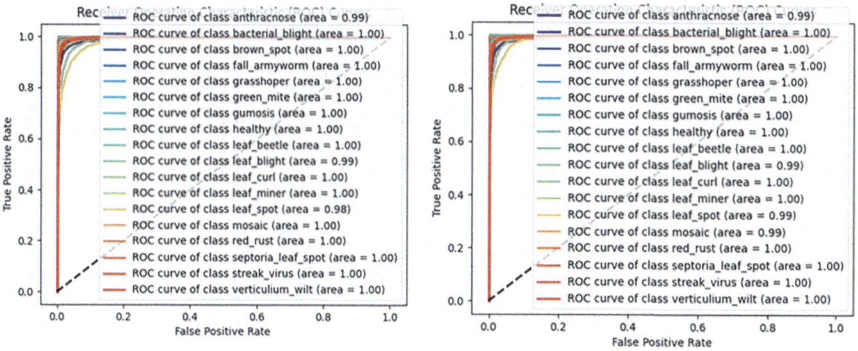


**Fig. 8** ROC curve for Models 1 and 2

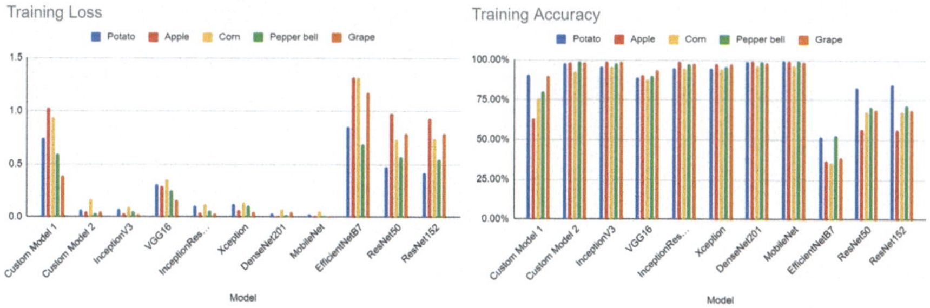


**Fig. 9** Training loss and accuracy across the models and crops

tion by a little margin. Regarding consistency, the two models' performance across all other classes stays constant, implying that they are both equally successful at distinguishing between the positive and negative classes across multiple thresholds.

## 3.7 *Transfer Learning Results*

This study delves into the field of transfer learning to assess the adaptability and performance of custom-developed convolutional neural networks. This method was used to broaden the models' application to a range of crops—potato, grape, apple, pepper bell, and corn—each with distinct imaging properties and disease presentations, therefore testing the models' adaptability and resilience in a larger agricultural environment. The performance data for each crop across the 11 models were meticulously documented during each stage of the machine learning process: training, validation, and testing. The findings are shown in Figs. 9 and 10.

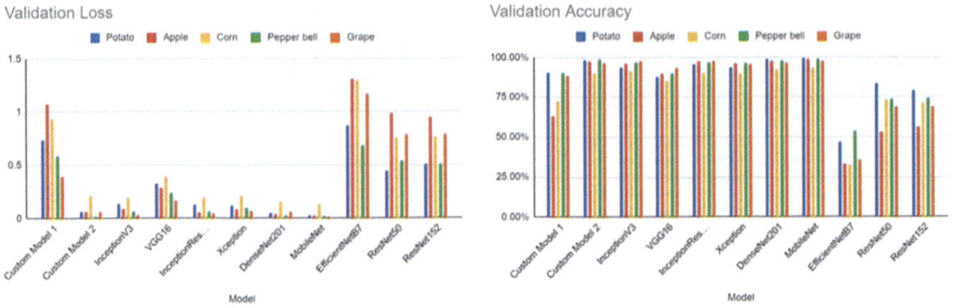


**Fig. 10** Validation loss and accuracy across models and crops

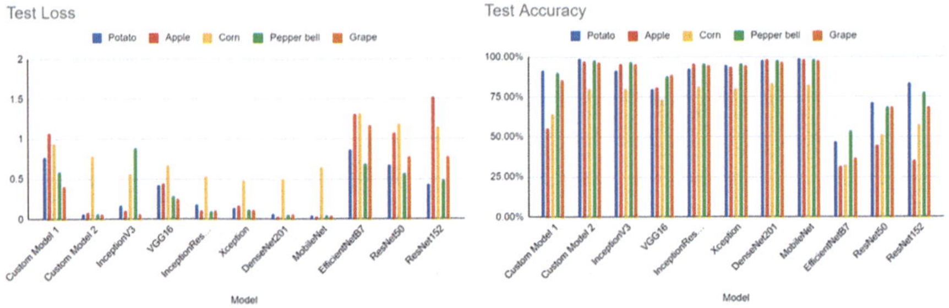


**Fig. 11** Testing loss and accuracy across models and crops

### 3.7.1 Potential Effects of the Differences

In a comparative investigation of transfer learning performance for custom Model 1 and custom Model 2 across a variety of crops, the results show a clear pattern in favour of custom Model 2. Custom Model 2 regularly beats custom Model 1 in criteria such as training loss, training accuracy, validation loss, validation accuracy, test loss, and test accuracy. Custom Model 2's outstanding performance is demonstrated by its high accuracy rates, which are almost flawless in many cases, as well as its ability to sustain these rates throughout the training and testing phases. This demonstrates not just a great fit to the training data but also an extraordinary capacity to generalize to new, previously unknown data.

Figure 11 shows that custom Model 1, although demonstrating respectable accuracy rates, particularly during the validation and test stages, indicates a level of efficacy in the models' capacity to learn and generalize. However, it shows a drop in test accuracy over various crops, indicating probable overfitting to the training data, which is unsuitable for a model that must work reliably across several datasets.

### 3.7.2 Observations

The overall trend noticed is that custom Model 2 has a strong predictive capability, retaining excellent performance even when presented with fresh data. This consistency is critical for practical applications that need the model to generate predictions in real-world circumstances outside of the training environment. The evidence strongly supports the usage of custom Model 2 in deployment situations because of its shown dependability and accuracy over a wide range of datasets.

The reason is that custom Model 2's higher performance in transfer learning tasks, despite its relative simplicity in comparison to the more sophisticated custom Model 1, provides a compelling rebuttal to the widely accepted view that model complexity corresponds directly with generalization capabilities. This result encourages a more comprehensive examination of the parameters impacting model performance when applied to fresh datasets.

Custom Model 1's architecture, while more complex, may not have been perfectly matched with the feature distributions of the transfer learning tasks. The intricacy of Model 1 likely resulted in a clever yet highly specialized feature extraction that succeeded in the first domain but lacked the flexibility to adapt to other domains. In contrast, custom Model 2's simpler design may have assisted the learning of more generalizable characteristics, which proved useful when the model met data from other distributions.

Furthermore, the effectiveness of regularization techniques cannot be underestimated. Custom Model 2 may have used more efficient regularization approaches to prevent overfitting and to maintain the model's flexibility. This flexibility is crucial in transfer learning, where the model must apply previously learned information to new challenges.

## 4 Conclusion

This chapter aimed to improve the use of deep learning in agricultural technology, namely for crop disease categorization. The adventure began with the creation of two unique convolutional neural networks (CNNs) designed to tackle the hard challenge of diagnosing various illnesses in crops using computer vision. The elimination of unnecessary layers and refining of the network topology resulted in a significant performance improvement. The training, validation, and testing steps produced a large amount of data, highlighting each model's strengths and limitations.

### 4.1 Insights and Implications

Discussion of findings allowed for interpretation and assessment of their relevance to agricultural technology in general. Model 2's improved performance demonstrated the value of model optimization. The successful implementation of transfer learning revealed the models' flexibility and scalability across several crops.

The consequences of this study are considerable. The study contributes to technological developments in precision agriculture by providing tools that have the potential to alter crop disease control, increase yields, and assure sustainable farming methods. This work provides a foundation for future research, promoting more exploration and innovation at the interface of computer vision and agriculture. As the research concludes, this study has been both demanding and gratifying. It has created new paths of investigation and established a precedent for the use of deep learning in agriculture. This work has enormous potential to affect not just the realm of agricultural technology but also the worldwide quest for food security and sustainable farming practices.

## References

1. A. Abbas, S. Jain, M. Gour, S. Vankudothu, Tomato plant disease detection using transfer learning with C-GAN synthetic images. Comput. Electron. Agric. **187**, 106279 (2021). https://doi.org/10.1016/j.compag.2021.106279
2. S. Albahli, M. Masood, Efficient attention-based CNN network (EANet) for multi-class maize crop disease classification. Front. Plant Sci. **13** (2022). https://www.frontiersin.org/journals/plant-science/articles/10.3389/fpls.2022.1003152
3. Ü. Atila, M. Uçar, K. Akyol, E. Uçar, Plant leaf disease classification using EfficientNet deep learning model. Eco. Inform. **61**, 101182 (2021). https://doi.org/10.1016/j.ecoinf.2020.101182
4. S. Bansal, A. Kumar, Automatic Deep Neural Network Hyper-Parameter Optimization for Maize Disease Detection. IOP Conf. Ser.: Mater. Sci. Eng. **1022**(1), 012089 (2021)
5. J. Basavaiah, A. Arlene Anthony, Tomato Leaf Disease Classification using Multiple Feature Extraction Techniques. Wirel. Pers. Commun. **115**(1), 633–651 (2020)
6. A. Bhujel, N.-E. Kim, E. Arulmozhi, J.K. Basak, H.-T. Kim, A Lightweight Attention-Based Convolutional Neural Networks for Tomato Leaf Disease Classification. Agriculture **12**(2), Article 2 (2022). https://doi.org/10.3390/agriculture12020228
7. Biological Tomato Leaf Disease Classification using Deep Learning Framework | Request PDF (n.d.). Retrieved 21 February 2024, from https://www.researchgate.net/publication/359522040_Biological_Tomato_Leaf_Disease_Classification_using_Deep_Learning_Framework
8. X. Chen, G. Zhou, A. Chen, J. Yi, W. Zhang, Y. Hu, Identification of tomato leaf diseases based on combination of ABCK-BWTR and B-ARNet. Comput. Electron. Agric. **178**, 105730 (2020)
9. F. Hamami, I.A. Dahlan, Classification of Tomato Disease using Convolutional Neural Network. IOP Conf. Ser.: Earth Environ. Sci **1038**(1), 012032 (2022). https://doi.org/10.1088/1755-1315/1038/1/012032
10. V. Jayaprakash, K. Akilesh kumar, M.S. Balamurugan, M.K. Rajagopal, *Cashew dataset generation using augmentation and RaLSGAN and a transfer learning based tinyML approach towards disease detection*. arXiv, arXiv:2304.08766 (2023). https://doi.org/10.48550/arXiv.2304.08766

11. P. Kanuparthi, V. Bejgam, V.M. Viswanatham, A Novel Approach of Ensembling the Transfer Learning Methods for Rice Plant Disease Detection and Classification. Webology **18**(2), 439–448 (2021)
12. S. Khan, M. Narvekar, Novel fusion of color balancing and superpixel based approach for detection of tomato plant diseases in natural complex environment. J. King Saud Univ. Comput. Inf. Sci. **34**(6, Part B), 3506–3516 (2022). https://doi.org/10.1016/j.jksuci.2020.09.006
13. J. Liu, X. Wang, Early recognition of tomato gray leaf spot disease based on MobileNetv2-YOLOv3 model. Plant Methods **16**(1), 83 (2020). https://doi.org/10.1186/s13007-020-00624-2
14. S. Mishra, R. Sachan, D. Rajpal, Deep Convolutional Neural Network based Detection System for Real-time Corn Plant Disease Recognition. Procedia Comput. Sci. **167**, 2003–2010 (2020). https://doi.org/10.1016/j.procs.2020.03.236
15. A. Nader, M.H. Khafagy, S.A. Hussien, Grape Leaves Diseases Classification using Ensemble Learning and Transfer Learning. Int. J. Adv. Comput. Sci. Appl. (IJACSA) **13**(7), Article 7 (2022)
16. S. Nandhini, K. Ashokkumar, Improved crossover based monarch butterfly optimization for tomato leaf disease classification using convolutional neural network. Multimed. Tools Appl. **80**(12), 18583–18610 (2021). https://doi.org/10.1007/s11042-021-10599-4
17. M. Nawaz, T. Nazir, A. Javed, M. Masood, J. Rashid, J. Kim, A. Hussain, A robust deep learning approach for tomato plant leaf disease localization and classification. Sci. Rep. ***12***(1), Article 1 (2022)
18. L.C. Ngugi, M. Abdelwahab, M. Abo-Zahhad, Tomato leaf segmentation algorithms for mobile phone applications using deep learning. Comput. Electron. Agric. **178**, 105788 (2020)
19. V. Ochango, G. Wambugu, J. Ndia, Feature Extraction using Histogram of Oriented Gradients for Image Classification in Maize Leaf Diseases. Int. J. Comput. Inf. Technol. (2279–0764) **11**(4) (2022)
20. G. Sambasivam, G.D. Opiyo, A predictive machine learning application in agriculture: Cassava disease detection and classification with imbalanced dataset using convolutional neural networks. Egypt. Inf. J. **22**(1), 27–34 (2021)
21. Sravan, V., Swaraj, K., Meenakshi, K., & Kora, P. (2021). WITHDRAWN: A deep learning based crop disease classification using transfer learning
22. L. Tan, J. Lu, H. Jiang, Tomato Leaf Diseases Classification Based on Leaf Images: A Comparison between Classical Machine Learning and Deep Learning Methods. AgriEngineering *3*(3), Article 3 (2021)
23. N. Zhang, H. Wu, H. Zhu, Y. Deng, X. Han, Tomato Disease Classification and Identification Method Based on Multimodal Fusion Deep Learning. Agriculture **12**(12), Article 12 (2022)

# Anomaly Detection in PD-NOMA Cognitive Radio IoT Networks

**Gilbert Machani, Simon Chege, and Tom Walingo**

## 1 Introduction

The use of power domain-non-orthogonal multiple access (PD-NOMA) in cognitive radio (CR) has demonstrated its efficiency in promoting spectrum use. Users are assigned different power levels and use the same resources, thus accommodating more users and ensuring efficient spectrum use [1]. However, malicious users (MUs) pose significant security risks to such systems by sending false occupancy reports, forcing unauthorized resource use and interfering with primary user (PU) signals. The most common types of attacks are the spectrum sensing data falsification (SSDF), primary user emulation (PUE) and jamming attacks [2]. SSDF is characterized by MUs, which send falsified spectrum sensing (SS) information that leads to wrong channel occupancy decision and denying legitimate users access to the spectrum resource. PUE attacks occur when a user changes its transmission characteristics to be like those of the PUs, thus leading to wrong sensing decision by the secondary user (SU) [3]. Jamming, on the other hand, entails users transmitting signals which interfere with the PU signals, thus disrupting normal operations and lowering the quality of service (QoS) [4]. An honest secondary user (SU) actively performs spectrum sensing (SS) and transmits data using the PU resources. MUs choose not to perform SS and interfere with normal network operations [5]. Several studies have shown the various techniques that can be used to monitor and identify MUs in cognitive radio.

G. Machani (✉) · S. Chege · T. Walingo
Electrical, Electronic & Computer Engineering, University of Kwazulu Natal, Durban, South Africa
e-mail: walingo@ukzn.ac.za



I. Woungang, S. K. Dhurandher (eds.), *The 7th International Conference on Wireless, Intelligent and Distributed Environment for Communication*,
Lecture Notes on Data Engineering and Communications Technologies 237,
https://doi.org/10.1007/978-3-031-80817-3_12

Several techniques have been suggested to deal with these attacks. In order to detect and defend the users against PUE attacks, ref. [6] proposed the Yardstick-based Threshold Allocation (YTA) scheme that leverages on the potential of energy level prediction to mitigate interference to PUs and safeguard legitimate SUs from MUs. The YTA works by assigning threshold values to the base station (BS) based on the distance between PUs and the distribution of MUs. Modern techniques involve the use of machine learning algorithms like support vector machine to identify and classify MUs in CR Internet of things (IoT) networks [7]. A neural network-based classification approach is used to handle PUE attacks by using signal characteristics to differentiate between PUs and attackers [8]. This technique is highly complex since it requires more computational resources for training and testing. A cost-benefit trade-off of byzantine attacks was analyzed in ref. [9] and an intrusion detection scheme was developed that uses the frequency property of the statistical data and a Markov chain data ordering technique. The model relies on the assumption of a rational behavior of MUs, which might not always be the case, and does not address an optimal and wider scope of attacks.

A technology that has great potential to reduce anomalies in CR is non-orthogonal multiple access (NOMA) due to the enhanced signal processing capability, layered security architecture, improved spectral efficiency, and dynamic resource allocation that it provides. The impact of sharing disruption on multi-carrier CR-NOMA is investigated in ref. [10] where the uplink transmission is jammed, thus denying SUs access to the spectrum. A disruption framework and strategy are proposed but the challenge still remains on how to reduce the effects of disruption on overall network performance. In ref. [11], the focus is on optimizing the system performance by factoring in the eavesdropper intercept and outage probabilities. The solution involved a machine learning-based continuous genetic algorithm, which presented serious complexity and interference limitations. The impact of eavesdropping on system reliability and security for energy-harvesting CR-NOMA is presented in ref. [12] where an optimal balance is sought to achieve the required secrecy rates. This solution is highly complex and does not address the competition between system security and reliability as transmit power increases. Power domain-NOMA (PD-NOMA) can separate multiple users based on their power levels, thus minimizing interference and quality of service (QoS) challenges. However, PD-NOMA CR has not been spared by attacks. According to ref. [13], NOMA CR brings in great advantage to the system but suffers from channel gain feedback falsification (CGFF) attack, which is addressed using an eigenvalue distribution detector.

Other techniques focus on the identification of any "out of the usual" characteristics of any user, herein termed as an anomaly, to identify MUs. In ref. [14], the authors presented an adaptive parameter-based singular spectrum analysis technique for anomaly detection and protection against interference. The proposed technique is highly complex and is not adaptive to parameter selection. In ref. [15], point anomaly and cumulative anomaly detection techniques using exponentially weighted moving average and anomaly factor, respectively, are analyzed. The authors noted the need for future optimization of the framework that is used to improve efficiency and performance. In an attempt to detect anomalies in the millimeter wave spectrum of CRN, ref. [16] compares the performance of deep learning (DL) models namely conditional generative adversarial network (C-GAN), auxiliary classifier generative adversarial network (AC-GAN), and variational auto encoder (VAE). These require a significant amount of time to train and test, which may not be applicable in real-time situations. The authors in ref. [17] applied the $k$-nearest neighbors (KNN) algorithm to improve detection in a cooperative spectrum sensing model where the PUs are considered as anomalies. This is highly dependent on $K$ (number of nearest neighbors). The Gaussian mixture model-based anomaly detection technique implemented in ref. [18] against byzantine attacks is computationally expensive and highly dependent on the characteristics of the data which compromise its performance in case the data do not exhibit Gaussian properties.

Since it is possible to identify variations in power allocation factors (PAFs) from spectrum data, we propose a simple technique that combines the partial periodic pattern mining (PPPM) and moving average time series (MATS) to detect anomalies in the network. Different from refs. [14–18], the proposed technique uses the $K$ out of $N$ rule for decision-making, which greatly reduces system complexity. The PPPM performs mining which aims at identifying frequent PAF behavior from the users' spectrum utilization. This works on historical data and helps the BS to identify long-term and periodic behavior of the users based on the PAF. The MATS works on real-time PAF data to identify irregular PAF patterns based on a sliding window.

## 2 System Model

The model in Fig. 1 considers an uplink PD-NOMA network where the BS communicates with a set of $M$ IoT user equipment ($UE$), including PUs, SUs and MUs, which perform PUE attacks. The power allocation factors (PAFs) observed or measured over time for the $m^{th}$ $UE$ is given by $\boldsymbol{\alpha}_m = \{\alpha_{m,t} | \, m \in UE, t \in T\}$, where $\alpha_{m,t}$ is the PAF of the $m^{th}$ $UE$ ($m = 1, \ldots, M$) at time $t$ and $T$ is the total duration for the observations. The BS acts as the fusion center (FC) to monitor and make decisions based on the observed data.

## 3 Anomaly Detection

In NOMA, the PAFs determine the transmission powers for different $UEs$ and helps to separate them to avoid interference [19]. PAFs which deviate from the known values might indicate that an MU is present. To monitor the PAF over time, we propose an anomaly detection technique that combines two techniques namely partial periodic pattern mining (PPPM) and the moving average time series (MATS) anomaly detection (Fig. 2).

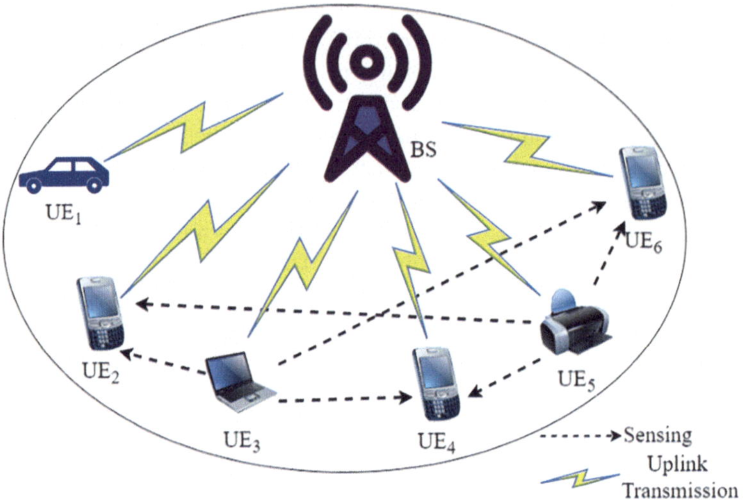


**Fig. 1** Uplink PD-NOMA-CRN-IoT anomaly detection model

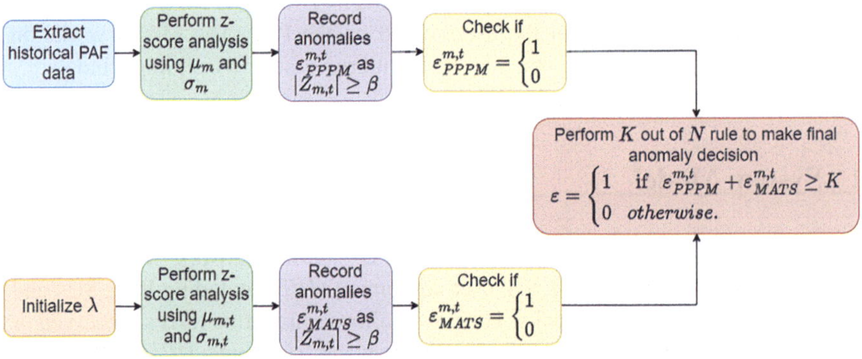


**Fig. 2** Architecture for the proposed anomaly detection technique

## *3.1 PAF Periodic Pattern Mining (PPPM)*

This works by analyzing historical spectral usage data of different *UEs* to identify regular and periodic PAF patterns [20]. Since PUE attack involves an MU emulating the PU signal to deny resources or ensure the SUs vacate the PU channel, PPPM can be utilized to check past PAF patterns and categorize the specific UE. The time series of the PAFs of the $m^{th}$ *UE*, i.e., $\boldsymbol{\alpha}_m = \{\alpha_{m,t}\}$, can be used to analyze the periodicity of the PAFs by using its discrete Fourier transform (DFT) given by

$$\hat{\alpha}_{m,k} = \sum_{t=0}^{C-1} \alpha_{m,t} e^{-j2\pi kt / C}, \tag{1}$$

given that $C$ is the number of periodic time components and $k = 0, 1, \ldots, C-1$. The $z$-score of the PAFs is computed as

$$Z_{m,t} = \frac{\alpha_{m,t} - \mu_m}{\sigma_m}, \tag{2}$$

where

$$\mu_m = \frac{1}{C} \sum_{t=0}^{C-1} \alpha_{m,t} \tag{3}$$

and

$$\sigma_m = \frac{1}{C} \sum_{t=0}^{C-1} \left(\alpha_{m,t} - \mu_m\right)^2. \tag{4}$$

The z-score obtained is compared with the set threshold ($\beta$) as

$$\left|Z_{m,t}\right| \geq \beta : \text{Anomaly detected}, \tag{5}$$

where $\beta$ is selected based on the fact that a majority of the selected points should lie closer to the mean. A value less than $\beta$ will result in more anomalies being detected, thus increasing the false positives (i.e., normal PAF flagged as anomalous). A greater value of $\beta$ will make the algorithm very strict thus making it miss some true anomalies [21]. The value of $\beta$ was selected and tested to balance these two contradicting concerns. The set of PAF anomalies are recorded as

$$\varepsilon_{\text{PPPM}}^{m,t} = \begin{cases} 1, \left|Z_{m,t}\right| \geq \beta \\ 0, \text{otherwise} \end{cases}. \tag{6}$$

## 3.2 *Moving Average Time Series (MATS)*

This technique performs the real-time continuous monitoring of the *UE* spectral usage to detect PAF variations that might be indicative of malicious activity. The moving average is calculated based on a sliding window ($\lambda$) from the current time of monitoring [22]. The moving PAF average ($\mu_{m,t}$) and standard deviation ($\sigma_{m,t}$) are calculated as

$$\mu_{m,t} = \frac{1}{\lambda} \sum_{l=t-\lambda+1}^{t} \alpha_{m,l} \tag{7}$$

and

$$\sigma_{m,t} = \frac{1}{\lambda} \sum_{l=t-\lambda+1}^{t} \left(\alpha_{m,l} - \mu_{m,t}\right)^2. \tag{8}$$

The $z$-score of the PAFs is computed as

$$Z_{m,t} = \frac{\alpha_{m,t} - \mu_{m,t}}{\sigma_{m,t}}. \tag{9}$$

The $z$-score obtained is compared with the set threshold ($\beta$) as in Eq. (5) and the set of PAF anomalies recorded as

$$\varepsilon_{\text{MATS}}^{m,t} = \begin{cases} 1, \left|Z_{m,t}\right| \geq \beta \\ 0, \text{otherwise} \end{cases}. \tag{10}$$

## 3.3 Proposed Technique

The proposed technique is a combination of PPPM and MATS for anomaly detection, which leverages on the historical PAF variations to explain current *UE* behavior. Figure 2 presents the architecture of the proposed technique. We formulate this as a $K$ out of $N$ rule, where the PPPM and MATS are regarded as voters in the decision-making process at the BS [23]. The overall anomaly indicator function ($\varepsilon$) of the $K$ out of $N$ rule is described as

$$\varepsilon = \begin{cases} 1, \varepsilon_{\text{PPPM}}^{m,t} + \varepsilon_{\text{MATS}}^{m,t} \geq K \\ 0, \text{otherwise} \end{cases}, \tag{11}$$

where $K$ is the minimum number of methods to indicate an anomaly and $N = 2$.

If $K = 1$,

$$\varepsilon = \begin{cases} 1, \text{if } \varepsilon_{\text{MATS}}^{m,t} = 1 \\ 0, \text{otherwise} \end{cases}. \tag{12}$$

If $K = 2$,

$$\varepsilon = \begin{cases} 1, \varepsilon_{PPPM}^{m,t} = 1 \text{ AND } \varepsilon_{\text{MATS}}^{m,t} = 1 \\ 0, \text{otherwise} \end{cases}. \tag{13}$$

## 4 Results and Discussion

The uplink performance of the proposed technique for PAF anomaly detection that combines PPPM and MATS with a $K$ out of $N$ rule-based decision-making technique is investigated. The data used consist of PAF recorded over a period of $T = 43$ days, where $\beta = 2$, $\lambda = 24$h and $C = 1$h. Six UEs were used whose PAFs are $\alpha_1 = 0.3$, $\alpha_2 = 0.4$, $\alpha_3 = 0.5$, $\alpha_4 = 0.6$, $\alpha_5 = 0.7$, $\alpha_6 = 0.8$. Any modulation scheme can be used for this system, and an assumption is made that an appropriate PAF detection or approximation scheme is available at the receiver. Figure 3 shows the variations of PAF over the period under consideration.

The variations indicate that the UEs utilize different PU PAFs at different times, except when it becomes malicious. As indicated in Fig. 4, the PPPM and MATS check for anomalies in the PAF data, which are indicated by the blue and orange dots respectively. The PPPM identifies points where the z-score, which is computed from the mean and standard deviation of the historical PAFs, exceeds $\beta$. The MATS calculates the z-score based on a rolling mean and identifies points which exceed $\beta$. This shows that both techniques can detect anomalous points in the PAF data. However, the anomalies detected do not always overlap, which indicates that each technique captures different aspects of the PAF data.

The $K$ out of $N$ rule provides a decision-making structure that combines the advantages of the individual techniques. For $K = 1$, an anomaly is detected if it has been detected by the MATS technique alone. This technique is highly sensitive to PAF deviations, thus, presenting a high number of anomalies detected. However, since only one technique is used for decision-making, the result of this matches that of the MATS technique and has higher likelihood for false positives. For $K = 2$, there is a reduction in the number of anomalies detected, thus highlighting the exclusive

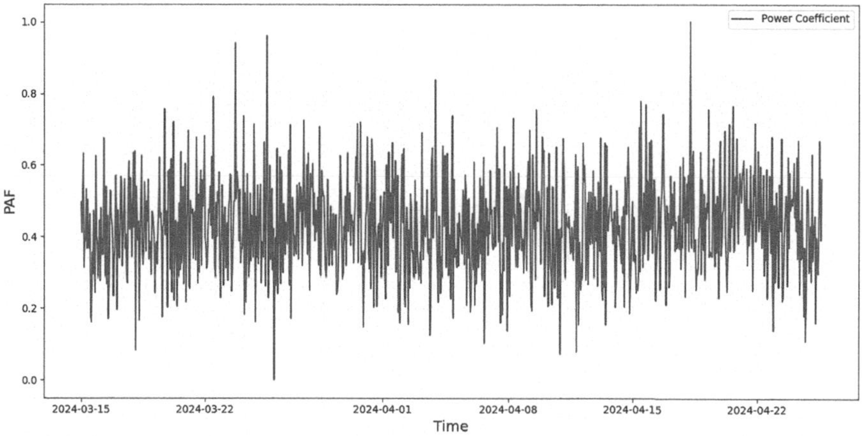


**Fig. 3** PAF variations

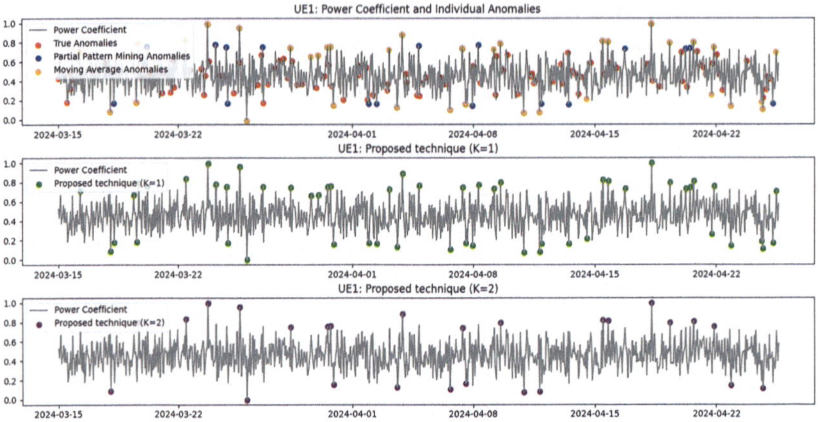


a) UE1 anomalies

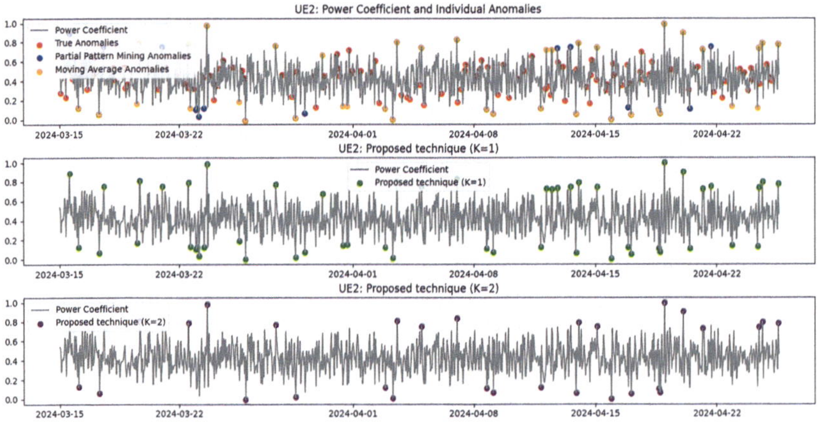


b) UE2 anomalies

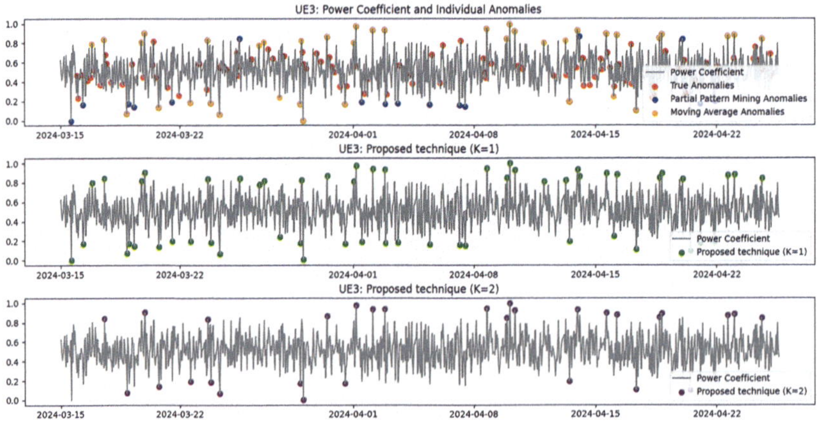


c) UE3 anomalies

**Fig. 4** (**a–f**) Anomaly detection performance for each UE

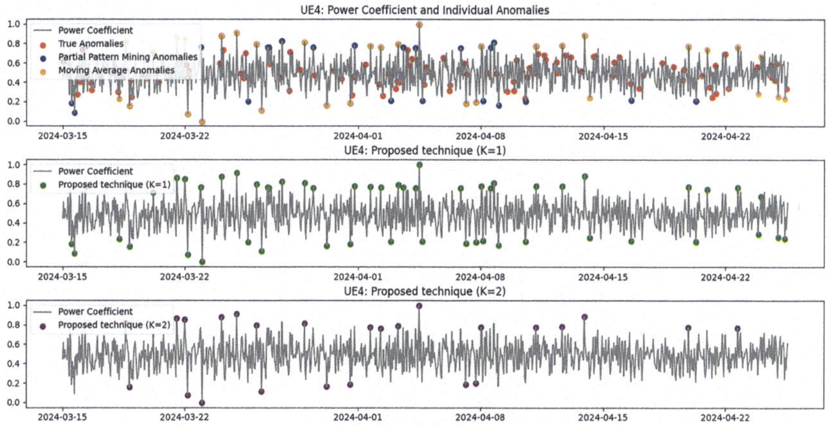


d) UE4 anomalies

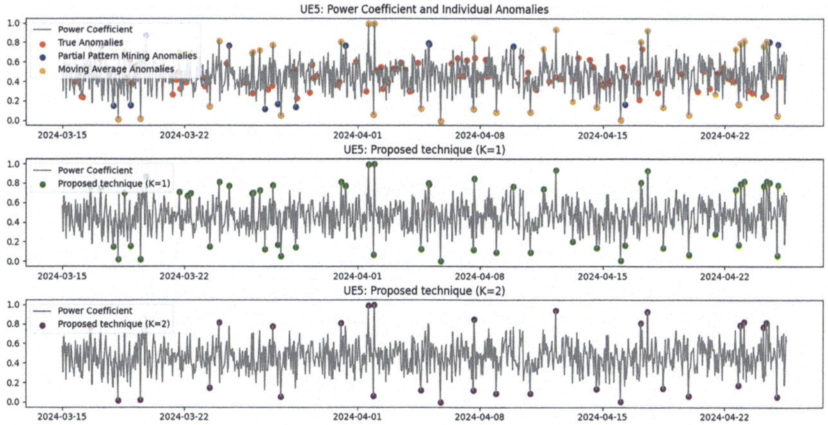


e) UE5 anomalies

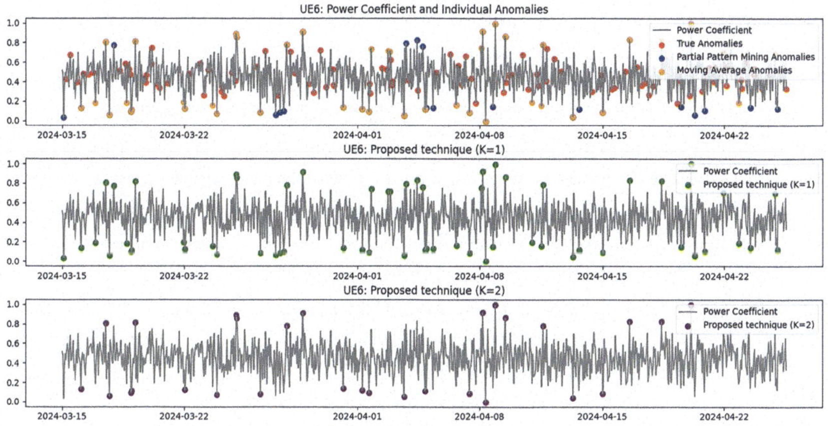


f) UE6 anomalies

**Fig. 4** (continued)

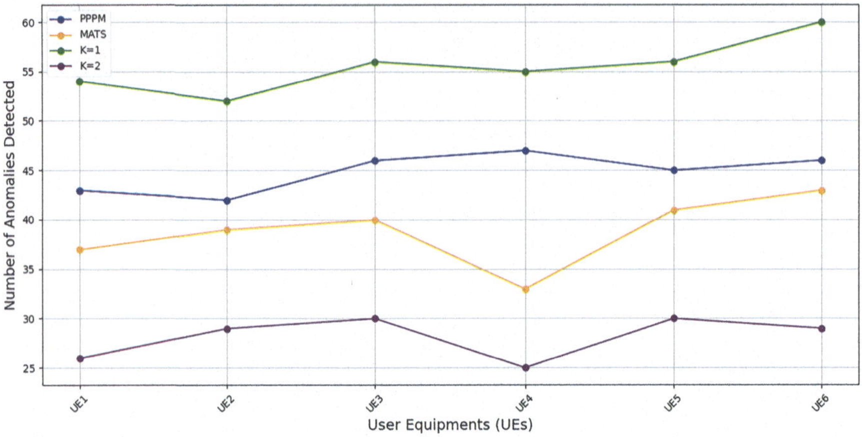


**Fig. 5** Variation in the number of anomalies detected with increase in PAF

nature, high confidence, and specificity that are brought about by seeking consensus between PPPM and MATS. However, the technique might overlook the anomalies detected by only one technique.

Regarding the anomaly detection performance while considering users with different PAFs, the proposed technique with $K = 1$ displays an increase in anomalies detected as PAF increases from the 1st to 6th UE. This suggests that an increase in PAF may result in more anomalies detected as shown in Fig. 5. It generally has the highest number of anomalies detected as compared to the other methods. The strict nature of $K = 2$ resulted in the lowest number of anomalies detected with increase in PAF. PPPM displays a slight increase in anomalies detected, while MATS detects more anomalies with increase in PAF.

The analysis was also done considering performance parameters like precision, false positive rate (FPR), recall, and F1 score. The result indicates that the proposed technique with $K = 2$ has the highest precision, meaning that it is the most effective in reducing the false positives, followed by $K = 1$, and PPPM and MATS have the lowest precision. A low FPR when $K = 2$ indicates that the technique has a very low chance of encountering false positives as compared to MATS and PPPM, while $K = 1$ has more false positives attributed to its aggressive anomaly detection nature. In terms of recall, $K = 2$ has the best performance, followed by $K = 1$ while PPPM and MATS have similar performance. This means that $K = 2$ can easily detect the real positives. The best balance between precision and recall is adequately achieved when $K = 2$, as seen by the high F1 score performance. A slightly lower F1 score is achieved for $K = 1$ as compared to $K = 2$, while PPPM has the lowest. Therefore, the proposed technique outperforms PPPM and MATS especially when $K = 2$ across all the performance metrics. This is summarized in Table 1.

**Table 1** Performance comparison

| | PPPM | MATS | Proposed ($K = 1$) | Proposed ($K = 2$) |
|---|---|---|---|---|
| Precision (%) | 25.6 | 5.1 | 33.3 | 35.9 |
| Recall (%) | 20.3 | 20.3 | 27.1 | 32.4 |
| F1 score (%) | 21.6 | 22.7 | 26.8 | 29 |
| FPR (%) | 28 | 19.9 | 34.8 | 17.4 |

## 5 Conclusion

In this chapter, an uplink PD-NOMA CRN IoT network is considered and the performance of two anomaly detection techniques is investigated. The individual techniques are first used to detect anomalies independently after which a flexible $K$ out of $N$ rule is used to assess the performance when the two techniques are combined. PPPM was used to identify deviations from the historical PAF data, while MATS was used to perform real-time anomaly detection by using a sliding window and mean. Simulation results indicate that combining the two techniques creates a balance between sensitivity, specificity, and confidence in anomaly detection evidenced by the precision and F1 score performance. The best performance is witnessed when $K = 2$. This emphasizes the robust nature of the proposed technique which can be used for reliable anomaly detection to inform the CRN of a possible malicious user to maintain the performance and integrity required. The proposed technique with $K = 1$ resulted in an increase in anomalies detected as PAF increased, while $K = 2$ had little correlation between the number of anomalies and PAF as evidenced by the slight increase in the anomalies detected with increase in PAF. The proposed technique has displayed better performance than the MATS and PPPM techniques. However, further research is recommended to further increase the performance of the proposed technique in PD-NOMA CRN.

**Acknowledgments** The authors appreciate the support of the Centre for Radio Access and Rural Technologies (CRART) at the University of KwaZulu Natal (UKZN) in producing this work.

## References

1. S. Chege, T. Walingo, Energy efficient resource allocation for uplink hybrid power domain sparse code nonorthogonal multiple access heterogeneous networks with statistical channel estimation. Trans. Emerg. Tel. Tech. **32**, e4185 (2021). https://doi.org/10.1002/ett.4185
2. G. Sharma, R. Sharma, A review on recent advances in spectrum sensing, energy efficiency and security threats in cognitive radio network, in *2015 International Conference on Microwave, Optical and Communication Engineering (ICMOCE)*, (Bhubaneswar, 2015), pp. 114–117. https://doi.org/10.1109/ICMOCE.2015.7489704

3. A.S. Hamood, S.B. Sadkhan, Cognitive radio network security status and challenges, in *2017 Annual Conference on New Trends in Information & Communications Technology Applications (NTICT)*, (Baghdad, 2017), pp. 1–6. https://doi.org/10.1109/NTICT.2017.7976105
4. H. Pirayesh, H. Zeng, Jamming attacks and anti-jamming strategies in wireless networks: A comprehensive survey. IEEE Commun Surv Tutor **24**(2), 767–809 (2022). https://doi.org/10.1109/COMST.2022.3159185
5. D.K. Jasim, S.B. Sadkhan, Cognitive radio network: Security and reliability trade-off - status, challenges, and future trend, in *2021 1st Babylon International Conference on Information Technology and Science (BICITS)*, (Babil, 2021), pp. 149–153. https://doi.org/10.1109/BICITS51482.2021.9509929
6. N. Sureka, K. Gunaseelan, Detection & Defense against primary user emulation attack in dynamic cognitive radio networks, in *2019 Fifth International Conference on Science Technology Engineering and Mathematics (ICONSTEM)*, (Chennai, 2019), pp. 505–510. https://doi.org/10.1109/ICONSTEM.2019.8918819
7. S. Hossain, S. Miah, Machine learning-based malicious user detection for reliable cooperative radio spectrum sensing in cognitive radio-internet of things. Machine Learning with Applications **5**, 100052 (2021). https://doi.org/10.1016/j.mlwa.2021.100052
8. V. Ponnusamy, K. Kottursamy, T. Karthick, B. Mukeshkrishnan, D. Malathi, T. Ahanger, Primary user emulation attack mitigation using neural network. Comput. Electr. Eng. **88**, 106849 (2020). https://doi.org/10.1016/j.compeleceng.2020.106849
9. J. Wu, Y. Yu, H. Zhu, T. Song, J. Hu, Cost-benefit tradeoff of byzantine attack in cooperative Spectrum sensing. IEEE Syst. J. **14**(2), 2532–2543 (2020). https://doi.org/10.1109/JSYST.2019.2952395
10. T.Y. Alkhamees, L.B. Milstein, Impact of sharing disruption in MC CR-NOMA. IEEE Access **11**, 82871–82881 (2023). https://doi.org/10.1109/ACCESS.2023.3300659
11. V.N. Vo et al., Outage probability minimization in secure NOMA cognitive radio systems with UAV relay: A machine learning approach. IEEE Trans. Cognit. Commun. Networking **9**(2), 435–451 (2023). https://doi.org/10.1109/TCCN.2022.3226184
12. P. Yan et al., EH cognitive network with NOMA: Perspective on impact of passive and active eavesdropping. IEEE Internet Things J. **11**(3), 5050–5062 (2024). https://doi.org/10.1109/JIOT.2023.3301631
13. S. Xia, X. Tao, N. Li, S. Wang, Malicious user detection in non-orthogonal multiple access based on Spectrum analysis. IEEE Signal Process Lett. **27**, 1390–1394 (2020). https://doi.org/10.1109/LSP.2020.3012826
14. Q. Dong, Z. Yang, Y. Chen, X. Li, K. Zeng, Anomaly detection in cognitive radio networks exploiting singular Spectrum analysis, in *Computer Network Security. MMM-ACNS 2017*, Lecture Notes in Computer Science(), ed. by J. Rak, J. Bay, I. Kotenko, L. Popyack, V. Skormin, K. Szczypiorski, vol. 10446, (Springer, Cham, 2017). https://doi.org/10.1007/978-3-319-65127-9_20
15. S.M.N. Uddin, N. Mansoor, M. Rahman, N. Mohammed, S. Hossain, A framework for event anomaly detection in cognitive radio based smart community, in *International Workshop on Computational Intelligence (IWCI)*, vol. 2016, (Dhaka, 2016), pp. 148–152. https://doi.org/10.1109/IWCI.2016.7860356
16. A. Toma, A. Krayani, L. Marcenaro, Y. Gao, C.S. Regazzoni, Deep learning for spectrum anomaly detection in cognitive mmWave radios, in *IEEE 31st Annual International Symposium on Personal, Indoor and Mobile Radio Communications*, vol. 2020, (London, 2020), pp. 1–7. https://doi.org/10.1109/PIMRC48278.2020.9217240
17. L. Lopez-Lopez, Á.G. Andrade, G. Galaviz, Cooperative spectrum sensing based on Anomaly detection and K nearest neighbors, in *2022 IEEE Future Networks World Forum (FNWF)*, vol. 2022, (Montreal), pp. 238–242. https://doi.org/10.1109/FNWF55208.2022.00049
18. A. Parmar, K. Shah, K.M. Captain, M. López-Benítez, J.R. Patel, Gaussian mixture model-based anomaly detection for defense against byzantine attack in cooperative Spectrum sensing. IEEE Trans. Cognit. Commun. Networking **10**(2), 499–509 (2024). https://doi.org/10.1109/TCCN.2023.3342409

19. S.M.R. Islam, N. Avazov, O.A. Dobre, K.-s. Kwak, Power-domain non-orthogonal multiple access (NOMA) in 5G systems: Potentials and challenges. IEEE Commun Surv Tutor **19**(2), 721–742 (2017). https://doi.org/10.1109/COMST.2016.2621116
20. P. Huang, C.-J. Liu, X. Yang, L. Xiao, J. Chen, Wireless Spectrum occupancy prediction based on partial periodic pattern mining. IEEE Trans Parallel Distrib Syst **25**(7), 1925–1934 (2014). https://doi.org/10.1109/TPDS.2013.283
21. R.S. Jebur, R.H. Thaher, Multiple FBG wavelength peaks detection using Z-score algorithm, in *2022 International Symposium on Multidisciplinary Studies and Innovative Technologies (ISMSIT)*, vol. 2022, (Ankara), pp. 223–226. https://doi.org/10.1109/ISMSIT56059.2022.9932700
22. A. Malki, E. Atlam, I. Gad, Machine learning approach of detecting anomalies and forecasting time-series of IoT devices. Alex. Eng. J. **61**(11), 8973–8986 (2022). https://doi.org/10.1016/j.aej.2022.02.038
23. A.A. Sharifi, M.J. Musevi Niya, Defense against SSDF attack in cognitive radio networks: Attack-aware collaborative Spectrum sensing approach. IEEE Commun. Lett. **20**(1), 93–96 (2016). https://doi.org/10.1109/LCOMM.2015.2499286

# Index




I. Woungang, S. K. Dhurandher (eds.), *The 7th International Conference on Wireless, Intelligent and Distributed Environment for Communication*, Lecture Notes on Data Engineering and Communications Technologies 237, https://doi.org/10.1007/978-3-031-80817-3

The manufacturer's authorised representative in the EU is Springer Nature Customer Service Centre GmbH, Europaplatz 3, 69115 Heidelberg, Germany. If you have any concerns regarding our products, please contact ProductSafety@springernature.com

Printed and bound by CPI Group (UK) Ltd, Croydon, CR0 4YY
01/10/2026
02225956-0004